NIST Technical Note 1814

Building Industry Reporting and Design for Sustainability (BIRDS)
Technical Manual and User Guide

Barbara C. Lippiatt
Joshua Kneifel
Priya Lavappa
Applied Economics Office
Engineering Laboratory

Sangwon Suh
Industrial Ecology Research Services, LLC

Anne Landfield Greig
Four Elements Consulting, LLC

September 2013

U.S. Department of Commerce
Penny Pritzker, Secretary

National Institute of Standards and Technology
Patrick D. Gallagher, Under Secretary of Commerce for Standards and Technology and Director

Abstract

Building stakeholders need practical metrics, data, and tools to support decisions related to sustainable building designs, technologies, standards, and codes. The Engineering Laboratory of the National Institute of Standards and Technology (NIST) has addressed this high priority national need by extending its metrics and tool for sustainable building products, known as Building for Environmental and Economic Sustainability (BEES), to whole buildings. Whole building sustainability metrics have been developed based on innovative extensions to life-cycle assessment (LCA) and life-cycle costing (LCC) approaches involving building energy simulations. The measurement system evaluates the sustainability of both the materials and the energy used by a building over time. It assesses the "carbon footprint" of buildings as well as 11 other environmental performance metrics, and integrates economic performance metrics to yield science-based measures of the business case for investment choices in high-performance green buildings.

Building Industry Reporting and Design for Sustainability (BIRDS) applies the new sustainability measurement system to an extensive whole building performance database NIST has compiled for this purpose. The BIRDS database includes energy, environmental, and cost measurements for 12 540 new commercial and non low-rise residential buildings, covering 11 building prototypes in 228 cities across all U.S. states for 9 study period lengths. The sustainability performance of buildings designed to meet current state energy codes can be compared to their performance when meeting four alternative building energy standard editions to determine the impact of energy efficiency on sustainability performance. The impact of the building location and the investor's time horizon on sustainability performance can also be measured.

Keywords

Building economics; economic analysis; life-cycle costing; life-cycle assessment; energy efficiency; commercial buildings

Preface

This study was conducted by the Applied Economics Office in the Engineering Laboratory (EL) at the National Institute of Standards and Technology (NIST). The study is designed to assess the energy, environmental, and cost impacts from the adoption of new state energy codes based on more stringent building energy standard editions. The intended audience is researchers and decision makers in the commercial building sector, and others interested in building sustainability.

Disclaimer

The policy of the National Institute of Standards and Technology is to use metric units in all of its published materials. Because this report is intended for the U.S. construction industry that uses U.S. customary units, it is more practical and less confusing to include U.S. customary units as well as metric units. Measurement values in this report are therefore stated in metric units first, followed by the corresponding values in U.S. customary units within parentheses.

Acknowledgements

The authors wish to thank all those who contributed ideas and suggestions for this report. They include Douglas Thomas and Dr. Robert Chapman of EL's Applied Economics Office, Dr. William Healy of EL's Energy and Environment Division, and Dr. Nicos S. Martys of EL's Materials and Structural Systems Division. A special thanks to the Industrial Ecology Research Services team of Shivira Tomar, Christine Chen, and Matthew Leighton for their superb technical support in developing whole building life cycle assessments for BIRDS. Also deserving thanks is Mr. Nicholas Long and the EnergyPlus Team for generating the initial energy simulations for this project. Thanks to Mr. Brian Presser for adapting the energy simulations to meet the study requirements and generating the final simulations used in the database. Thanks to Mr. Nathaniel Soares for developing the initial version of the BIRDS database. Finally, the many Beta testers of BIRDS deserve a special thanks for contributing suggestions leading to substantial improvements in the tool.

Author Information

Joshua Kneifel
Economist
National Institute of Standards and Technology
Engineering Laboratory
100 Bureau Drive, Mailstop 8603
Gaithersburg, MD 20899-8603
Tel.: 301-975-6857
Email: joshua.kneifel@nist.gov

Priya Lavappa
Computer Specialist
National Institute of Standards and Technology
Engineering Laboratory
100 Bureau Drive, Mailstop 8603
Gaithersburg, MD 20899-8603
Tel.: 301-975-4522
Email: priya.lavappa@nist.gov

Contents

List of Tables

List of Figures

1 Introduction

1.1 Purpose

Building stakeholders need practical metrics, data, and tools to support decisions related to sustainable building designs, technologies, standards, and codes. The Engineering Laboratory of the National Institute of Standards and Technology (NIST) has addressed this high priority national need by extending its metrics and tool for sustainable building products, known as Building for Environmental and Economic Sustainability (BEES), to whole buildings. Whole building sustainability metrics have been developed based on innovative extensions to life-cycle assessment (LCA) and life-cycle costing (LCC) approaches involving building energy simulations. The measurement system evaluates the sustainability of both the materials and the energy used by a building over time. It assesses the "carbon footprint" of buildings as well as 11 other environmental performance metrics, and integrates economic performance metrics to yield science-based measures of the business case for investment choices in high-performance green buildings.

Building Industry Reporting and Design for Sustainability (BIRDS) applies the new sustainability measurement system to an extensive whole building performance database NIST has compiled for this purpose. The BIRDS database includes energy, environmental, and cost measurements for 12 540 new commercial and non low-rise residential buildings, covering 11 building prototypes in 228 cities across all U.S. states for 9 study period lengths. The sustainability performance of buildings designed to meet current state energy codes can be compared to their performance when meeting four alternative building energy standard editions to determine the impact of energy efficiency on sustainability performance. The impact of the building location and the investor's time horizon on sustainability performance can also be measured.

The new idea is to address building sustainability measurement in a holistic, integrated manner that considers complex interactions among building materials, energy technologies, and systems across dimensions of performance, scale, and time. The energy, environment, and cost data in BIRDS measure building operating energy use through detailed energy simulations, building materials use through innovative life-cycle material inventories, and building costs over time.

1.2 Background

A wave of interest in sustainability gathered momentum in 1992 with the Rio Earth Summit, during which the international community agreed upon a definition of sustainability in the Bruntland report: "meeting the needs of the present generation without compromising the ability of future generations to meet their own needs." (Brundtland, 1987) In the context of sustainable development, needs can be thought to include the often-conflicting goals of environmental quality, economic well-being, and social justice. While the intent of the 1992 summit was to initiate environmental and social progress, it seemed to have instead brought about greater debate over the inherent conflict between sustainability and economic development.

This conflict is particularly apparent within the construction industry. Frequently, well-intentioned green development plans are not executed for economic reasons, and economic development plans fail to materialize over concerns for the environment and public health. Thus, an integrated approach to sustainable construction—one that simultaneously considers both environmental and economic performance—lies at the heart of reconciling the conflict. For this reason, the BIRDS approach considers both the environmental and economic dimensions of sustainability. BIRDS does not consider the social dimension of sustainability due to its current lack of rigorous measurement methods.

2 BIRDS Approach

2.1 Rethink Sustainability Measurement

One standardized and preferred approach for scientifically measuring the environmental performance of industrial products and systems is life cycle assessment (LCA). LCA is a "cradle-to-grave," systems approach for measuring environmental performance. The approach is based on two principles. First, the belief that all stages in the life of a product generate environmental impacts and must be analyzed, including raw materials acquisition, product manufacture, transportation, installation, operation and maintenance, and ultimately recycling and waste management. An analysis that excludes any of these stages is limited because it ignores the full range of upstream and downstream impacts of stage-specific processes. LCA broadens the environmental discussion by accounting for shifts of environmental problems from one life-cycle stage to another. The second principle is that multiple environmental impacts must be considered over these life-cycle stages in order to implement a trade-off analysis that achieves a genuine reduction in overall environmental impact, rather than a simple shift of impact. By considering a range of environmental impacts, LCA accounts for problem-shifting from one environmental medium (land, air, water) to another.

The LCA method is typically applied to products, or simple product assemblies, in a "bottom up" manner. The environmental inputs and outputs to all the production processes throughout a product's life cycle are compiled. These product life cycle "inventories" quantify hundreds, even thousands, of environmental inputs and outputs. This is a data-intensive, time-consuming, and expensive process that must be repeated for every product.

The bottom-up approach becomes unwieldy and cost prohibitive for complex systems, such as buildings, that involve potentially hundreds of products. Furthermore, a building's sustainability is not limited to the collective sustainability of its products. The manner in which designers integrate these products and systems at the whole building level has a large influence on another major dimension of its sustainability performance, operating energy use.

The many dimensions of whole building environmental performance are ultimately balanced against its economic performance. While a 2006 poll by the American Institute of Architects showed that 90 % of U.S. consumers would be willing to pay more to reduce their home's environmental impact, they would pay only $4 000 to $5 000, or about 2 %, more.[1] Even the most environmentally conscious policymaker or building designer will ultimately weigh environmental benefits against economic costs. To satisfy their stakeholders, the green building community needs to promote and design buildings with an attractive balance of environmental and economic performance.

---

[1] January 2006 survey cited in *Washington Post*, 8/6/06, p M3 (Green Buildings article by Sacha Cohen).

These considerations require a different way of thinking about sustainability performance for buildings. In the BIRDS model, a unifying LCA framework developed for the U.S. economy is applied to the U.S. construction sector and its constituent building types. Through this "top-down" LCA approach, a series of baseline sustainability measurements are made for prototypical buildings, yielding a common yardstick for measuring sustainability with roots in well-established national environmental and economic statistics. Using detailed "bottom-up" data compiled through traditional LCA approaches, the baseline measurements for prototypical buildings are then "hybridized" to reflect a range of improvements in building energy efficiency, enabling assessment of their energy, environmental, and economic benefits and costs. The idea is to provide a cohesive database and measurement system based on sound science that can be used to prioritize green building issues and to track progress over time as design and policy solutions are implemented.

The BIRDS hybrid LCA approach combines the advantages of both bottom-up and top-down approaches—namely the use of higher-resolution, bottom-up data and the use of regularly-updated, top-down statistical data without truncation (Suh et al., 2004; Suh and Huppes, 2005). The hybrid approach generally reduces the uncertainty of existing pure bottom up or pure top down systems: it helps reduce truncation error in the former and increases the resolution of the latter (Suh et al., 2004).

Operating energy use—a key input to whole building LCAs—is assessed in BIRDS using the bottom-up approach. Energy use is highly dependent upon a building's function, size, location, and the efficiency of its energy technologies. Energy efficiency requirements in current energy codes for commercial buildings vary across states, and many states have not yet adopted the newest energy standard editions. As of December 2011, state energy code adoptions range across all editions of the *American Society of Heating, Refrigerating and Air-Conditioning Engineers Energy Standard for Buildings except Low-Rise Residential Buildings* (ASHRAE-90.1-1999, -2001, -2004, and -2007). Some states do not have a code requirement for energy efficiency, leaving it up to the locality or jurisdiction to set its own requirement. To address these issues, operating energy use in BIRDS is tailored to commercial and non low-rise residential building types, locations, and energy codes. The BIRDS database includes operating energy use predicted though energy simulation of 5 alternative building designs for 11 building types in 228 U.S. locations, with each design complying with some version of the energy code (90.1) or a higher-performing "Low Energy Case" building design based on *ASHRAE 189.1-2009* (*Standard for the Design of High-Performance Green Buildings Except Low-Rise Residential Buildings*).

Like operating energy use, a building's economic performance is dependent upon a building's function, location, and the efficiency of its energy technologies. Construction material and labor costs vary by building type and location, as do maintenance, repair, and replacement costs over time. Energy technologies for compliance with a given ASHRAE energy code edition vary across U.S. climate zones, as do their costs. Finally, a building's operating energy costs vary according to the quantity and price of energy use, which depend upon the building's location and fluctuate over time. All these variables are accounted for in the BIRDS database, as shown in Figure 2-1.

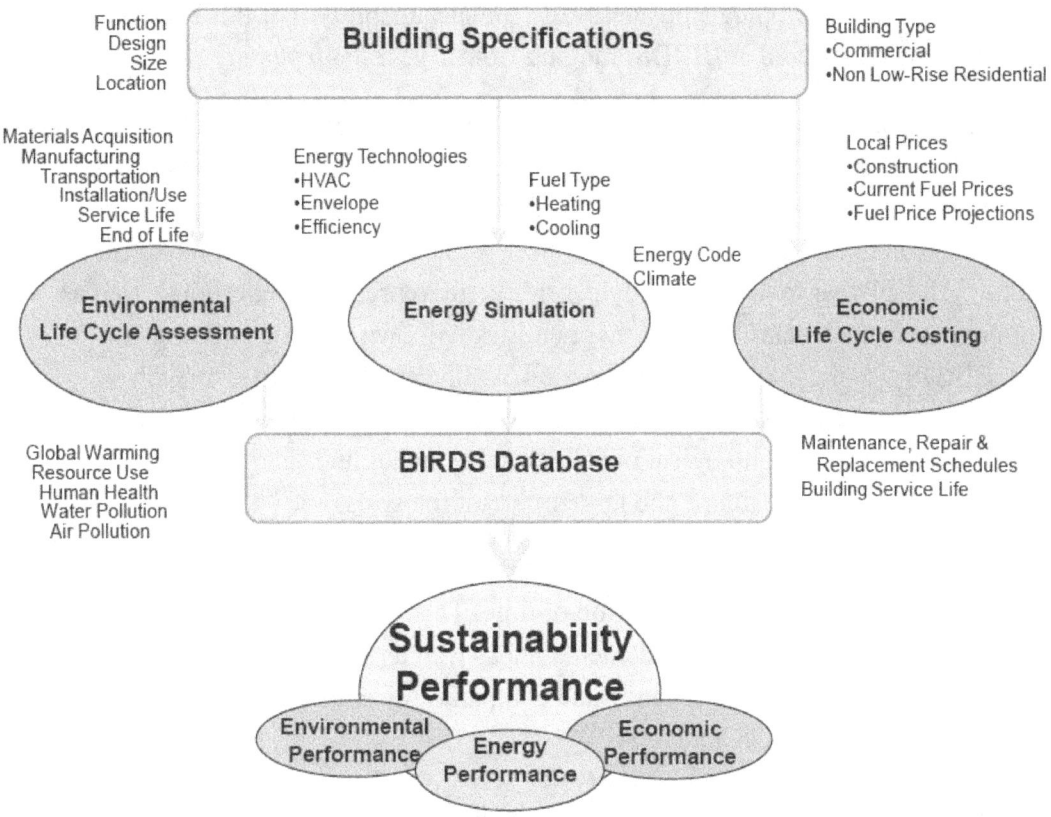

Figure 2-1 BIRDS Overview

2.2 Establish Consistency

This new way of measuring building sustainability performance requires that special attention be paid to establishing consistency among its many dimensions. While BIRDS develops separate performance metrics for building energy, environmental, and economic performance, they are all developed using the same parameters and assumptions. For each of the 12 540 buildings included in the BIRDS database, consistent design specifications are used to estimate its operating energy use, environmental life-cycle impacts, and life-cycle costs. The building energy simulation, for example, specifies the same building envelope and HVAC technologies as do the bottom-up energy technology LCAs and the cost estimates.

One of the most important dimensions requiring BIRDS modeling consistency is the study period. The study period is the number of years of building operation over which energy, environmental, and economic performance are assessed. In economic terms, the study period represents the investor's time horizon. Over what period of time are the environmental and economic costs and benefits related to the capital investment decision of interest to the investor or policymaker? Since different stakeholders have different time perspectives, there is no one correct study period for developing a business case for sustainability. For this reason, 9 different study period lengths are offered in BIRDS, ranging from 1 year to 40 years.

Nine study period lengths are chosen to represent the wide cross section of potential investment time horizons. A 1-year study period is representative of a developer that intends to sell a property soon after it is constructed. A 5-year to 15-year study period best represents a building owner's time horizon because few owners are concerned about costs realized beyond a decade into the future. The 20-year to 40-year study periods better represents institutions, such as colleges or government agencies, because these entities will own or lease buildings for 20 or more years. BIRDS sets the maximum study period at 40 years for consistency with requirements for federal building life-cycle cost analysis (Energy Independence and Security Act of 2007). Beyond 40 years, technological obsolescence becomes an issue, data become too uncertain, and the farther in the future, the less important the costs.

Once the BIRDS user sets the length of the study period, the energy, environmental, and economic data are all normalized to that period of time. This involves adjustments to a building's operating, maintenance, repair, and replacement data as well as to its remaining value at the end of the study period. This assures consistency and comparability among the three metrics, and is one of the strengths of the BIRDS approach.

The next 3 chapters go into more detail regarding the modeling of energy, environmental, and economic performance in BIRDS.

3 Energy Performance Measurement

The operating energy component of the BIRDS database was built following the framework developed in Kneifel (2010) and further expanded in Kneifel (2011a) and Kneifel (2011b). It includes the results of 12 540 whole building energy simulations covering 5 energy efficiency designs for 11 new commercial building types, 228 cities across the United States, and 9 study period lengths.

3.1 Building Types

The building characteristics in Table 3-1 describe the 11 building types included in BIRDS, which include 2 dormitories, 2 apartment buildings, a hotel, 3 office buildings, a school, a retail store, and a restaurant. These building types represent 46 % of the existing U.S. commercial building stock floor space.[2] The prototype buildings range in size from 465 m^2 (5000 ft^2) to 41 806 m^2 (450 000 ft^2). The building abbreviations defined in Table 3-1 are used to represent the building types in tables throughout this report.

Table 3-1 Building Characteristics

Building Type	Bldg. Abbr.	Floors	Floor Height m (ft)	Wall	Roof†	Pct. Glazing	Building Size m^2 (ft^2)	Occupancy Type	U.S. Floor Space (%)
Dormitory	DORMI04	4	3.66 (12)	Mass	IEAD	20 %	3097 (33 333)	Lodging	7 1 %
Dormitory	DORMI06	6	3.66 (12)	Steel	IEAD	20 %	7897 (85 000)		
Hotel	HOTEL15	15	3.05 (10)	Steel	IEAD	100 %	41 806 (450 000)		
Apartment	APART04	4	3.05 (10)	Mass	IEAD	12 %	2787 (30 000)		
Apartment	APART06	6	3.15 (10)	Steel	IEAD	14 %	5574 (60 000)		
School, High	HIGHS02	2	4.57 (15)	Mass	IEAD	25 %	12 077 (130 000)	Education	13.8 %
Office	OFFIC03	3	3.66 (12)	Mass	IEAD	20 %	1858 (20 000)	Office	17.0 %
Office	OFFIC08	8	3.66 (12)	Mass	IEAD	20 %	7432 (80 000)		
Office	OFFIC16	16	3.05 (10)	Steel	IEAD	100 %	24 155 (260 000)		
Retail Store	RETAIL1	1	4.27 (14)	Mass	IEAD	10 %	743 (8000)	Mercantile*	6.0 %
Restaurant	RSTRNT1	1	3.66 (12)	Wood	IEAD	30 %	465 (5000)	Food Service	2 3 %

*Only includes non-mall floor area.
†IEAD = Insulation Entirely Above Deck

3.2 Building Designs

Current state energy codes are based on different editions of the *International Energy Conservation Code (IECC)* or *ASHRAE 90.1 Standard*, which have requirements that vary based on a building's characteristics and the climate zone of the location. For the BIRDS database, the *ASHRAE 90.1 Standard*-equivalent design is used to meet current state energy codes and to define the alternative building designs. Additionally, a "Low Energy Case" building design based on *ASHRAE 189.1-2009*, and which goes beyond *ASHRAE 90.1-2007* requirements, is

---

[2] Based on the Commercial Building Energy Consumption Survey (CBECS) database

included as a building design alternative. For simplicity, this design may be referred to as an "edition" of the energy standard throughout the remainder of this report.

Table 3-2 shows that commercial building energy codes as of December 2011 vary by state. In a few instances, local jurisdictions have adopted energy standard editions that are more stringent than the state energy codes.[3] These cities are also included in Table 3-2.

Table 3-2 Energy Code by State and City Exception

Location	Energy Code	Location	Energy Code	Location	Energy Code
AK	None	IN	2007	NV	2004
AL	None	KS	None	NY	2007
Huntsville	2001	KY	2007	OH	2007
AR	2001	LA	2007	OK	None
AZ	None	MA	2007	OR	2007
Flagstaff	2004	MD	2007	PA	2007
Phoenix	2004	ME	None	RI	2007
Tucson	2004	MI	2007	SC	2004
CA	2007	MN	2004	SD	None
CO	2001	MO	None	Huron	2001
Grand Junction	2004	St Louis	2001	TN	2004
CT	2007	MS	None	TX	2007
DE	2007	MT	2007	UT	2007
FL	2007	NC	2007	VA	2007
GA	2007	ND	None	VT	2007
HI	2004	NE	2007	WA	2007
IA	2007	NH	2007	WI	2007
ID	2007	NJ	2007	WV	2001
IL	2007	NM	2007	WY	None

Note: Some city ordinances require energy codes that exceed state energy codes.
Note: State codes as of December 1, 2011.

State energy codes vary from *ASHRAE 90.1-1999* to *ASHRAE 90.1-2007* with some regional trends shown in Figure 3-1. The states in the central U.S. tend to wait longer to adopt newer *ASHRAE 90.1 Standard* editions. However, there are many cases in which energy codes of neighboring states vary drastically. For example, Missouri has no state energy code while of the 8 surrounding states, 2 have no state energy code, 1 has adopted *ASHRAE 90.1-2001*, 1 has adopted *ASHRAE 90.1-2004*, and 4 have adopted *ASHRAE 90.1-2007*.

---

[3] Local and jurisdictional requirements are obtained from the Database of State Incentives for Renewables and Efficiency (DSIRE). State energy code requirements targeting only public buildings and green standards are ignored.

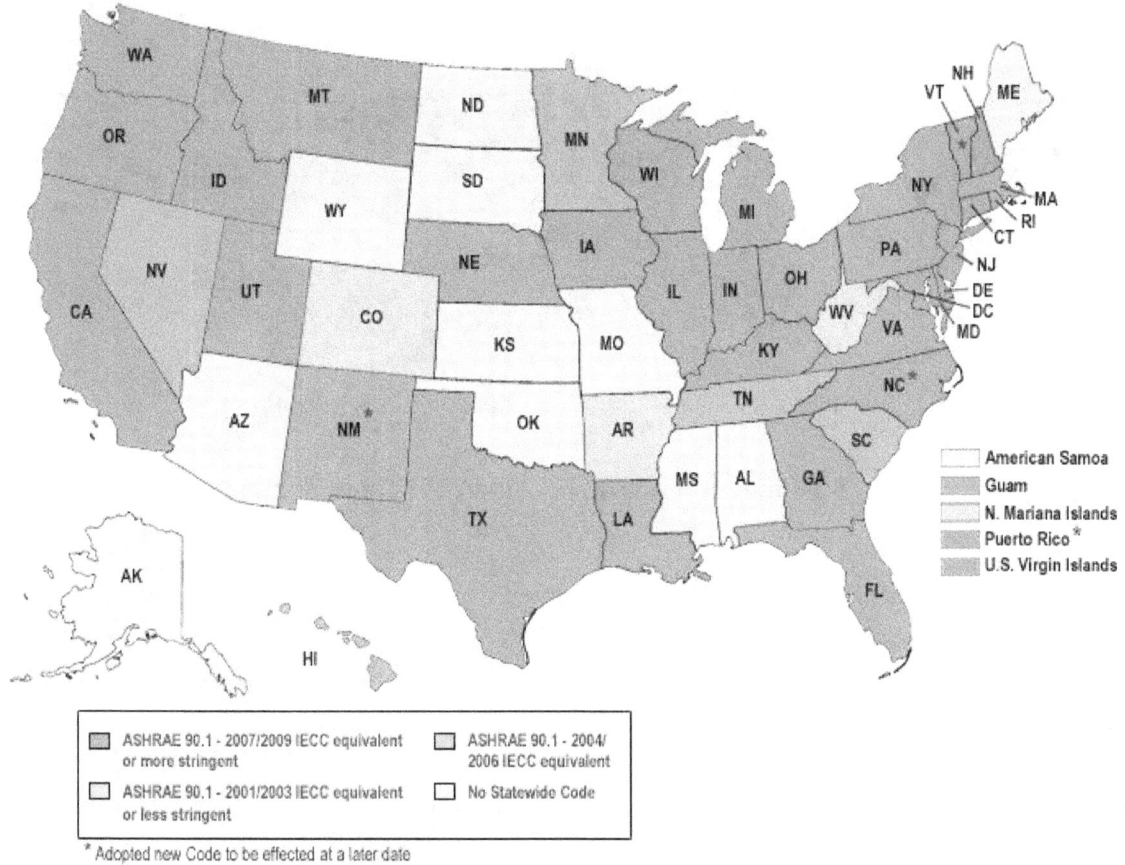

Figure 3-1 State Commercial Energy Codes[4]

The prototype buildings are designed to meet the requirements for each of the editions of *ASHRAE 90.1 (-1999, -2001, -2004,* and *-2007*) and an additional building design option defined as the "Low Energy Case" (LEC). The LEC design is based on *ASHRAE 189.1-2009* (*Standard for the Design of High-Performance Green Buildings Except Low-Rise Residential Buildings*) and goes beyond *ASHRAE 90.1-2007* in a number of ways. The LEC design increases the thermal efficiency of insulation and windows beyond *ASHRAE 90.1-2007*, reduces the lighting power density, and adds daylighting and window overhangs. The LEC design assumes the same HVAC equipment efficiency as required by *ASHRAE 90.1-2007*. For BIRDS, *ASHRAE 90.1-1999* is assumed to be "common practice," and is used for the building design requirements in states with no statewide energy code.

The 228 cities and current *ASHRAE* climate zones for the United States are mapped in Figure 3-2. These cities are selected for three reasons. First, the cities are spread out to represent the entire United States, and represent as many climate zones in each state as possible. Second, the locations cover all the major population centers in the country. Third, multiple locations for a climate zone within a state are included to allow building costs to vary for each building design.

---

[4] Figure was obtained from the DOE Building Technologies Program in December 2011.

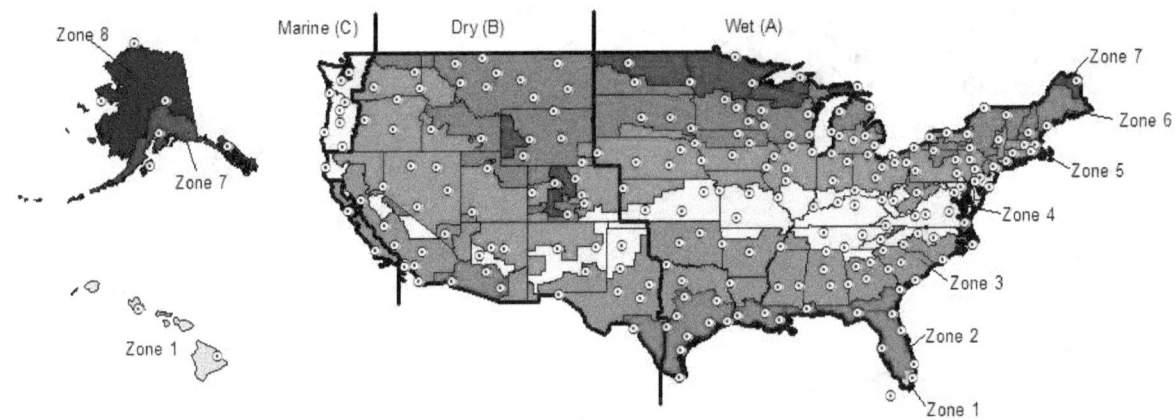

Figure 3-2 Cities and *ASHRAE-2004/2007* Climate Zones

The *ASHRAE*-defined climate zones were consolidated from the 26 climate zones shown in Table 3-3 for *ASHRAE 90.1-1999* and *90.1-2001* into 8 climate zones in *ASHRAE 90.1-2004* and *90.1-2007*. The consolidation is more complex than simply grouping the *90.1-2001* climate zones together. The zones in Table 3-3 are based on a city's cooling degree days at a base of 10 °C (50 °F) (CDD50) and heating degree days at a base of 18 °C (65 °F) (HDD65) while the zones in Figure 3-2 are based on a county's cooling and heating degree days. The generalization results in some cities, located in the same climate zone in *ASHRAE 90.1-2001,* being located in a different climate zone in *ASHRAE 90.1-2004*. The new climate zones are further separated into subzones, "wet," "dry," and "marine," as shown in Figure 3-2, for a total of 16 subzones.

Table 3-3 *ASHRAE 90.1-1999/2001* Climate Zone Definitions

CDD50	Climate Zone											
10801+	1											
9001-10800	2											
7201-9000	3	5										
5401-7200	4	6	8	10								
3601-5400		7	9	11	13	16						
1801-3600				12	14	17	19	21				
0-1800					15	18	20	22	23	24	25	26
HDD65	0-900	901-1800	1801-2700	2701-3600	3601-5400	5401-7200	7201-9000	9001-10800	10801-12600	12601-16200	16201-19800	19801+

CDD50 = Annual Cooling Degree Days base 10 C (50 F)
HDD65 = Annual Heating Degree Days base 18 C (65 F)

Occupancy types for the prototype buildings are based on the 1999 Commercial Building Energy Consumption Survey (CBECS) "principal building activity" categories. The principal building activity determines the maximum people density, electrical plug density, and lighting power density, and the correlating density schedules. Table 3-4 summarizes each building type's maximum densities. The maximum number of occupants varies between 1 person per 7.0 m^2 (75 ft^2) to 1 person per 27.9 m^2 (300 ft^2). The maximum interior lighting density and daily schedule vary based on the occupancy type and the edition of *ASHRAE 90.1* or Low Energy Case (LEC) building design. All building types are assumed to have zero external lighting loads. The maximum electrical equipment load varies between 1.08 W/m^2 (0.10 W/ft^2) to 8.07 W/m^2 (0.75 W/ft^2).

Table 3-4 Occupancy, Lighting, and Equipment

Bldg. Abbr.	ASHRAE 90.1 Occupancy Type	Max. Occ.	m^2 (ft^2) Per Occupant	Lighting W/m^2 (W/ft^2)	Equipment W/m^2 (W/ft^2)
DORMI04	Dormitory	99	23.2 (250)	8.6 to 16 1 (0.8 to 1.5)	2.69 (0.25)
DORMI06	Dormitory	342	23.2 (250)	8.6 to 16 1 (0.8 to 1.5)	2.69 (0.25)
HOTEL15	Hotel	1800	23.2 (250)	8.6 to 18 3 (0.8 to 1.7)	2.69 (0.25)
APART04	Dormitory	90	23.2 (250)	8.6 to 18 3 (0.8 to 1.7)	2.69 (0.25)
APART06	Dormitory	240	23.2 (250)	8.6 to 18 3 (0.8 to 1.7)	2.69 (0.25)
ELEMS01	School	602	7.0 (75)	10.8 to 16.1 (1.0 to 1.5)	5.38 (0.50)
HIGHS02	School	1740	7.0 (75)	10.8 to 16.1 (1.0 to 1.5)	5.38 (0.50)
OFFIC03	Office	72	25.5 (275)	8.6 to 14.0 (0.8 to 1.3)	8.07 (0.75)
OFFIC08	Office	288	25.5 (275)	8.6 to 14.0 (0.8 to 1.3)	8.07 (0.75)
OFFIC16	Office	944	25.5 (275)	8.6 to 14.0 (0.8 to 1.3)	8.07 (0.75)
RETAIL1	Retail	27	27.9 (300)	16.1 to 20.5 (1.5 to 1.9)	2.69 (0.25)
RSTRNT1	Dining: Fast Food	50	9.3 (100)	14.0 to 19.4 (1.3 to 1.8)	1.08 (0.10)

The square footage, number of floors, floor height, wall type, roof type, percent glazing, and heating, ventilation, and air conditioning (HVAC) system for each building type are based on the RS Means *Square Foot Cost Estimator* (*SFCE*) default prototype specifications. The cooling system is assumed to run on electricity while the heating system is assumed to run on natural gas. Table 3-5 identifies the heating and cooling equipment for each building type.

Table 3-5 HVAC Equipment by Building Type

Building Type	Cooling Equipment	Heating Equipment
DORMI04	Rooftop Packaged Unit	Furnace
DORMI06	Air-Cooled Chiller	Hot Water Boiler
HOTEL15	Water-Cooled Chiller	Hot Water Boiler
APART04	Air-Cooled Chiller	Hot Water Boiler
APART06	Air-Cooled Chiller	Hot Water Boiler
HIGHS02	Water-Cooled Chiller	Hot Water Boiler
OFFIC03	Rooftop Packaged Unit	Furnace
OFFIC08	Rooftop Packaged Unit	Furnace
OFFIC16	Water-Cooled Chiller	Hot Water Boiler
RETAIL1	Rooftop Packaged Unit	Furnace
RSTRNT1	Rooftop Packaged Unit	Furnace

Table 3-6 shows how air infiltration and mechanical ventilation rates vary by building type. Infiltration rates range between 0.3 air changes per hour (ACH) to 0.6 ACH. Minimum mechanical ventilation rates range between 0.4 ACH to 1.3 ACH. Total minimum ACH varies between 0.7 ACH to 1.9 ACH.

Table 3-6 Air Infiltration and Mechanical Ventilation

Bldg. Abbr.	Building Size m^2 (ft^2)	Floor Height m (ft)	CBECS Occupancy Type	Infiltration (ACH)	Ventilation (ACH)	Total ACH
DORMI04	3097 (33 333)	3.66 (12)	Lodging	0.3	0.4	0.7
DORMI06	7432 (80 000)	3.66 (12)		0.3	0.4	0.7
HOTEL15	41 806 (450 000)	3.05 (10)		0.3	0.5	0.8
APART04	2787 (30 000)	3.05 (10)		0.3	0.5	0.8
APART06	5574 (60 000)	3.15 (10)		0.3	0.5	0.8
HIGHS02	12 077 (130 000)	4.57 (15)	Education	0.3	1.0	1.3
OFFIC03	1858 (20 000)	3.66 (12)	Office	0.3	0.4	0.7
OFFIC08	7432 (80 000)	3.66 (12)		0.3	0.4	0.7
OFFIC16	24 155 (260 000)	3.05 (10)		0.3	0.5	0.8
RETAIL1	743 (8000)	4.27 (14)	Mercantile	0.4	0.6	1.0
RSTRNT1	465 (5000)	3.66 (12)	Food Service	0.6	1.3	1.9

3.3 Energy Simulation Design

Operating energy use is simulated for the 12 540 BIRDS buildings using the U.S. Department of Energy's *EnergyPlus Example File Generator (EEFG)*. The *EEFG* narrows down a building's description into simple, high-level characteristics: building geometry, orientation, number of floors, floor height, building type, wall and roof construction type, window-to-wall ratio, and location. The remaining building parameters are defined based on the chosen building energy standard and various default values based on the building type.

The original simulations obtained from the *EEFG* assumed a unitary HVAC system with gas heat for all building types. The HVAC system in the simulation was replaced with a system that best represents the system defined in the commercial prototype buildings in the cost database, the RSMeans *CostWorks Square Foot Cost Estimator (SFCE)*. The cooling and heating systems defined in the RSMeans CostWorks SFCE are specified in Table 5-3 of the Economic Performance chapter.

The efficiency of a particular piece of HVAC equipment is determined by the selected building design, which in BIRDS equates to the selected edition of *ASHRAE 90.1*. The building envelope design determines the capacity necessary for the HVAC equipment to meet the thermal load, which in turn requires the simulation of the design days. Based on the HVAC equipment capacity and the HVAC equipment type, the *ASHRAE 90.1* efficiency requirement reported in Table 5-3 was determined.

The HVAC systems are automatically sized for each location by EnergyPlus based on three design day outdoor conditions that are more restrictive than those recommended in the ASHRAE Fundamentals Handbook. The cooling load is based on two sets of design conditions based on the Typical Meteorological Year (TMY2) data: 0.4 % design dry-bulb temperature and mean coincident wet bulb temperature, and 0.4 % design wet-bulb temperature and mean coincident dry bulb temperature. The heating load is based on the 99.6 % dry-bulb design conditions. Both the heating and cooling auto-sizing use a sizing factor of 1.2.

The simulations assume parameter values for the exterior envelope that represent the performance of each surface as a single material. For example, a window is represented as a single layer with parameter values that represent the combined performance characteristics of each layer of the window. The individual components of the window (e.g., panes, coatings, films, gas fill, etc.) are not specified in the simulation, only the overall U-factor, Solar Heat Gain Coefficient (SHGC), and Visual Transmittance (VT) of the window.

Following are the details of each building type summarized above.

3.3.1   4-Story Apartment Building
The 4-story apartment building has mass walls, insulation entirely above the roof deck, operable windows, and a window-to-wall ratio of 12 %.[5] The detailed assumptions used in the energy simulations are described below.

Building Envelope
The energy efficiency characteristics of the building envelope are determined by the building's location and the edition of *ASHRAE 90.1*. The window characteristics (U-factor, SHGC, and VT) are based on the *ASHRAE 90.1* requirements for operable windows for 10.1 % to 20.0 % glazing. The wall and roof efficiency characteristics are based on the *ASHRAE 90.1* requirements for

---

[5] Window to wall ratio is defined as the percentage of the exterior wall area represented by windows.

residential buildings with above grade, mass wall construction and insulation entirely above the roof deck.

Heating, Ventilation, and Air Conditioning

There are four main aspects to the heating, ventilation, and air conditioning of a building: equipment, operating conditions, air infiltration, and mechanical ventilation. The HVAC equipment is a packaged electric air-cooled chiller and natural gas-fired hot water boiler. Each building type that falls into the "Lodging" CBECS category has the same constant heating and cooling setpoint temperatures shown in Figure 3-3: 21 °C (70 °F) for heating and 24 °C (76 °F) for cooling.

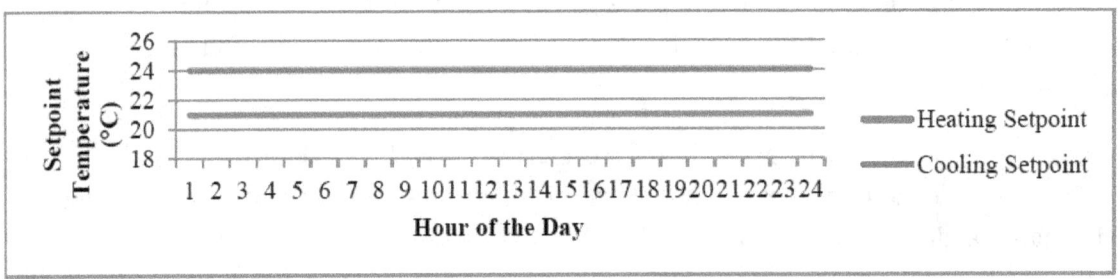

Figure 3-3 APART04 Setpoint Temperature Schedules

Air infiltration and mechanical ventilation are assumed to be constant across all editions of *ASHRAE 90.1*, with an infiltration rate of 0.177 m^3/s per floor (0.30 ACH) and minimum mechanical ventilation of 0.284 m^3/s per floor (0.48 ACH).

Occupancy, Lighting, and Electrical Loads

The peak occupancy for the 4-story apartment building is assumed to be 120 people or 1 person per 23.2 m^2 (250 ft^2). The schedule in Figure 3-4 shows that the greatest occupancy occurs (i.e., highest fraction of peak) over the nighttime while the lowest occupancy is during the middle of the day.

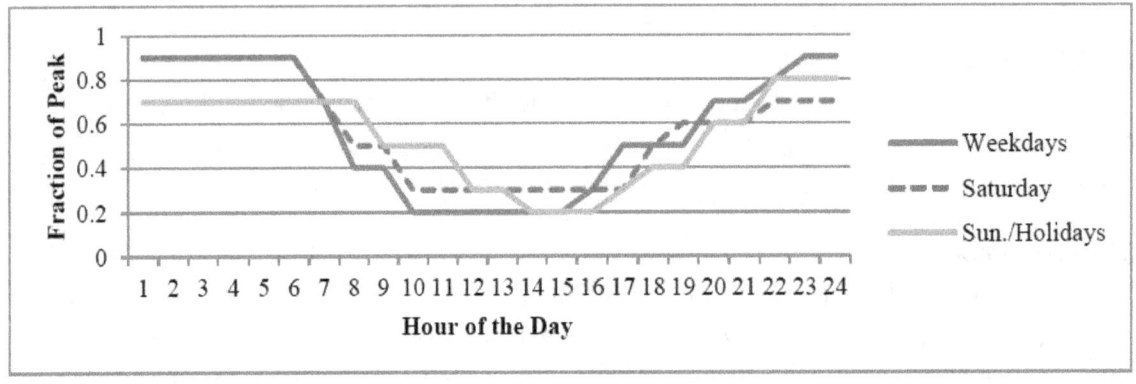

Figure 3-4 APART04 Occupancy Schedule

The energy simulation assumes between 8.6 W/m² (0.8 W/ft²) and 18.3 W/m² (1.7 W/ft²) of lighting density depending on the building design (e.g., edition of *ASHRAE 90.1* or LEC). The lighting load schedules, as a fraction of peak lighting loads, in Figure 3-5 are representative of typical residential occupant activity where the greatest loads are in the late evening between 7:00 PM and 11:00 PM. There is also a spike in lighting loads in the morning between 7:00 AM and 10:00 AM. The lighting loads also vary slightly based on the day of the week.

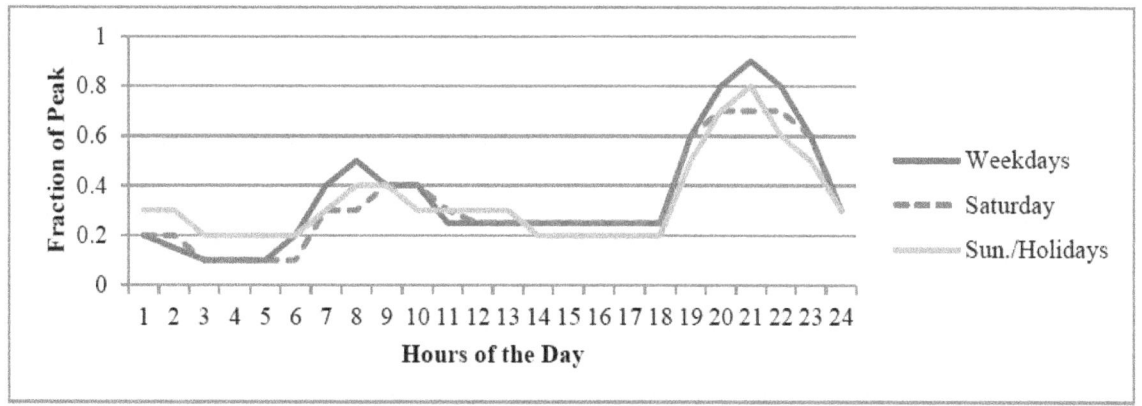

Figure 3-5 APART04 Lighting Schedule

The peak electrical equipment load is 7500 W, or 2.69 W/m² (0.25 W/ft²). Similar to lighting loads, the electrical load schedule in Figure 3-6 is highly correlated with occupant activity.

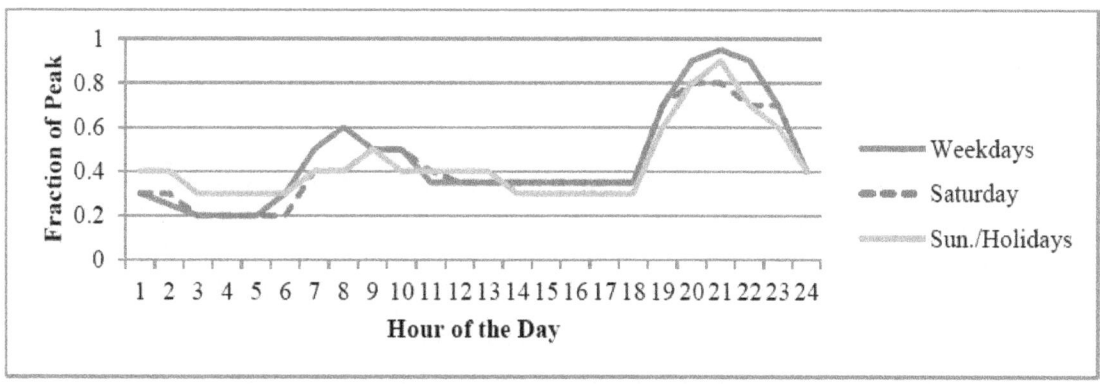

Figure 3-6 APART04 Electrical Load Schedule

3.3.2   6-Story Apartment Building

The 6-story apartment building has mass walls, insulation entirely above the roof deck, operable windows, and a window-to-wall ratio of 14 %. The detailed assumptions used in the energy simulations are described below.

Building Envelope

The energy efficiency characteristics of the building envelope are determined by the building's location and the edition of *ASHRAE 90.1*. The window characteristics (U-factor, SHGC, and VT) are based on the *ASHRAE 90.1* requirements for operable windows for 10.1 % to 20.0 % glazing.

The wall and roof efficiency characteristics are based on the *ASHRAE 90.1* requirements for residential buildings with above grade, mass wall construction and insulation entirely above the roof deck.

Heating, Ventilation, and Air Conditioning

There are four main aspects to the heating, ventilation, and air conditioning of a building: equipment, operating conditions, air infiltration, and mechanical ventilation. The HVAC equipment is a packaged electric air-cooled chiller and natural gas-fired hot water boiler. Each building type that falls into the "Lodging" CBECS category has the same constant heating and cooling setpoint temperatures shown in Figure 3-7: 21 °C (70 °F) for heating and 24 °C (76 °F) for cooling.

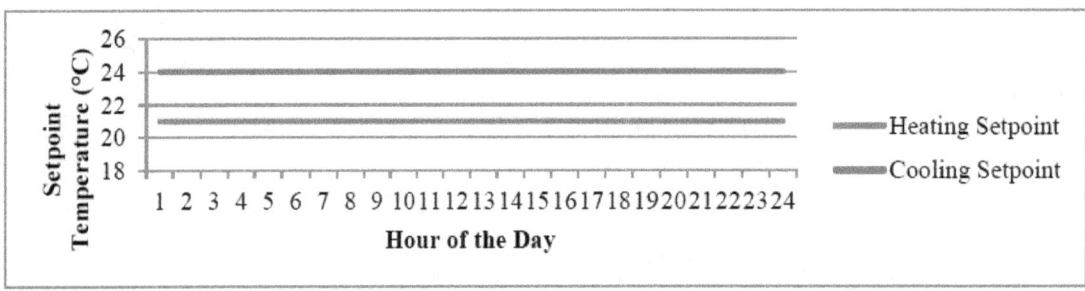

Figure 3-7 APART06 Setpoint Temperature Schedules

Air infiltration and mechanical ventilation are assumed to be constant across all editions of *ASHRAE 90.1*, with an infiltration rate of 0.244 m³/s per floor (0.30 ACH) and minimum mechanical ventilation of 0.379 m³/s per floor (0.47 ACH).

Occupancy, Lighting, and Electrical Loads

The peak occupancy for the 6-story apartment building is assumed to be 240 people or 1 person per 23.2 m² (250 ft²). The schedule in Figure 3-8 shows that the greatest occupancy occurs over the nighttime while the lowest occupancy is during the middle of the day.

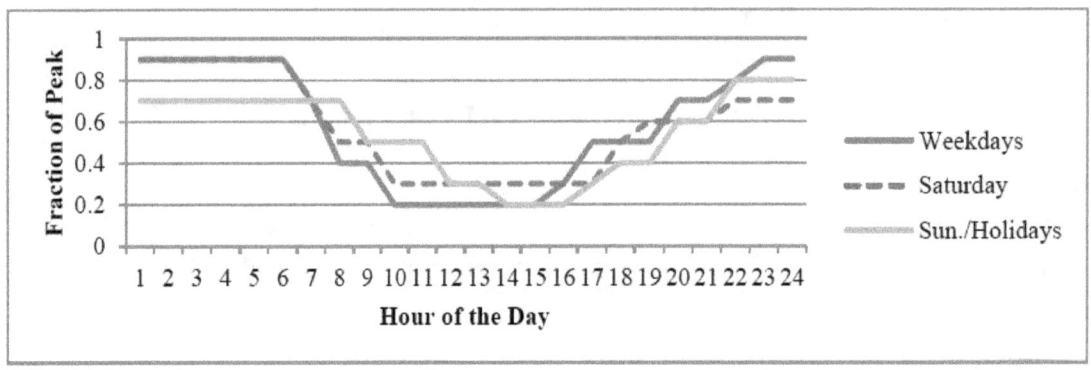

Figure 3-8 APART06 Occupancy Schedule

The energy simulation assumes between 8.6 W/m^2 (0.8 W/ft^2) and 18.3 W/m^2 (1.7 W/ft^2) of lighting density depending on the building design (e.g., edition of *ASHRAE 90.1* or LEC). The lighting load schedules, as a fraction of peak lighting loads, in Figure 3-9 are representative of typical residential occupant activity where the greatest loads are in the late evening between 7:00 PM and 11:00 PM. There is also a spike in lighting loads in the morning between 7:00 AM and 10:00 AM. The lighting loads also vary slightly based on the day of the week.

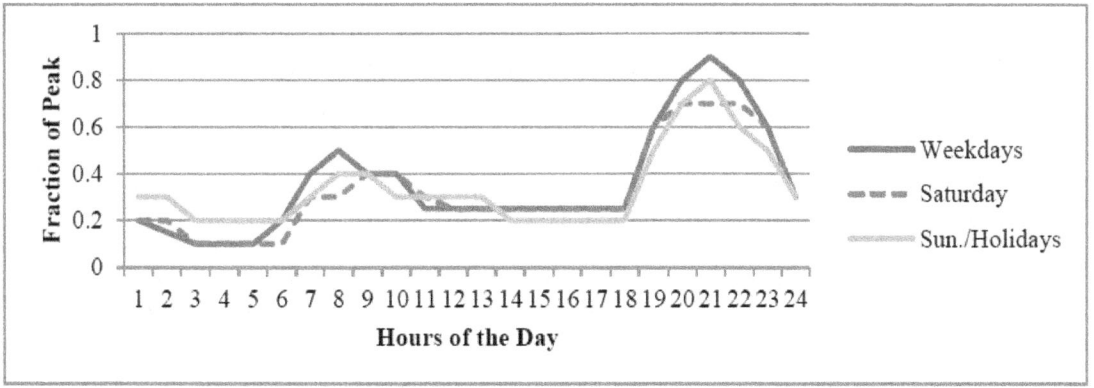

Figure 3-9 APART06 Lighting Schedule

The peak electrical equipment load is 5625 W, or 2.69 W/m^2 (0.25 W/ft^2). Similar to lighting loads, the electrical load schedule in Figure 3-10 is highly correlated with occupant activity.

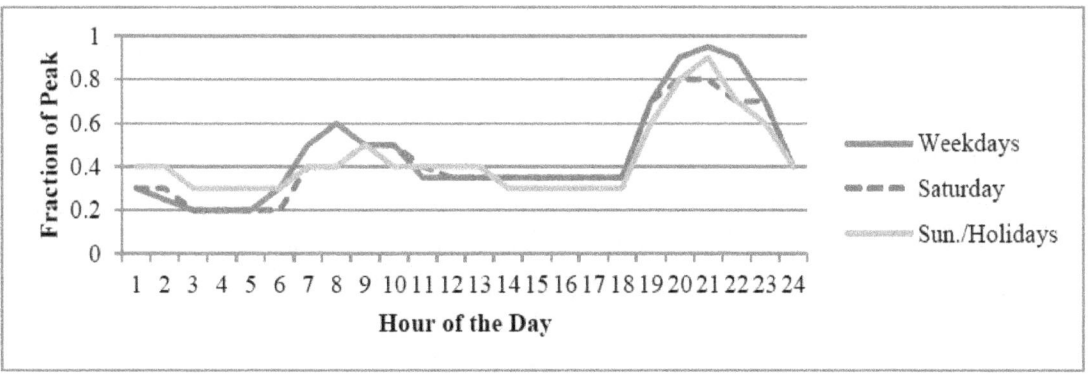

Figure 3-10 APART06 Electrical Load Schedule

3.3.3 4-Story Dormitory
The 4-story dormitory has mass walls, insulation entirely above the roof deck, operable windows, and a window-to-wall ratio of 20 %. The detailed assumptions used in the energy simulations are described below.

Building Envelope
The energy efficiency characteristics of the building envelope are determined by the building's location and the edition of *ASHRAE 90.1*. The window characteristics (U-factor, SHGC, and VT) are based on the *ASHRAE 90.1* requirements for operable windows for 10.1 % to 20.0 % glazing.

The wall and roof efficiency characteristics are based on the *ASHRAE 90.1* requirements for residential buildings with above grade, mass wall construction and insulation entirely above the roof deck.

Heating, Ventilation, and Air Conditioning

There are four main aspects to the heating, ventilation, and air conditioning of a building: equipment, operating conditions, air infiltration, and mechanical ventilation. The HVAC equipment is a packaged electric air-cooled chiller and natural gas-fired hot water boiler. Each building type that falls into the "Lodging" CBECS category has the same constant heating and cooling setpoint temperatures shown in Figure 3-11: 21 °C (70 °F) for heating and 24 °C (76 °F) for cooling.

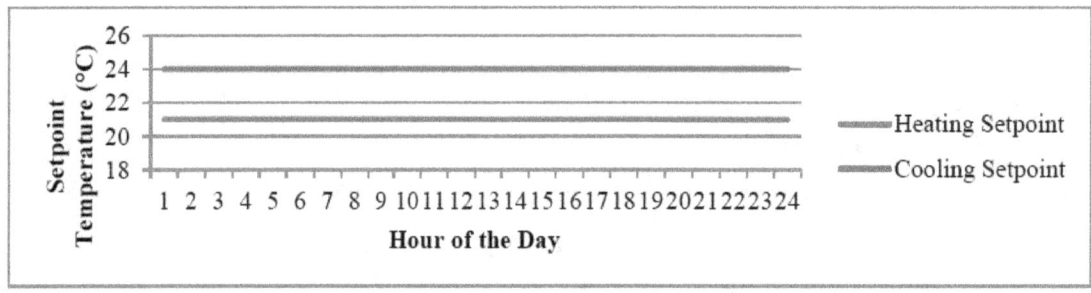

Figure 3-11 DORMI04 Setpoint Temperature Schedules

Air infiltration and mechanical ventilation are assumed to be constant across all editions of *ASHRAE 90.1*, with an infiltration rate of 0.236 m^3/s per floor (0.30 ACH) and minimum mechanical ventilation of 0.316 m^3/s per floor (0.40 ACH).

Occupancy, Lighting, and Electrical Loads

The peak occupancy for the 3-story dormitory is assumed to be 132 people or 1 person per 23.2 m^2 (250 ft^2). The schedule in Figure 3-12 shows that the greatest occupancy occurs over the nighttime while the lowest occupancy is during the middle of the day.

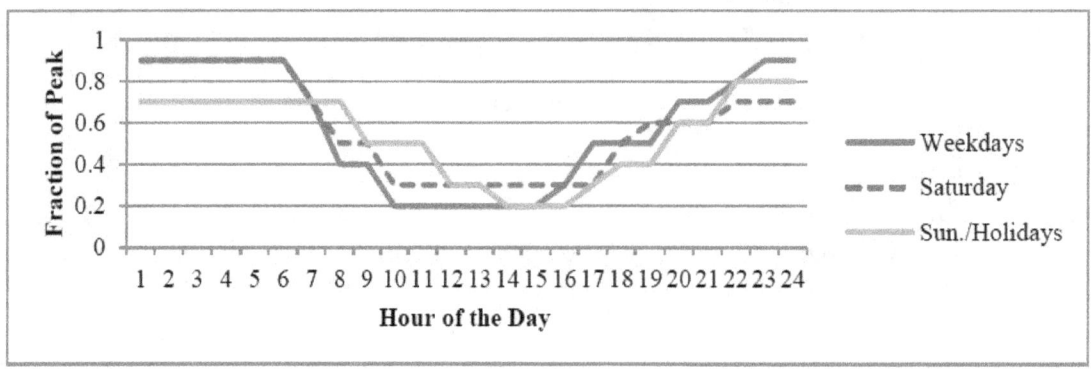

Figure 3-12 DORMI04 Occupancy Schedule

18

The energy simulation assumes between 8.6 W/m^2 (0.8 W/ft^2) and 18.3 W/m^2 (1.7 W/ft^2) of lighting density depending on the building design (e.g., edition of *ASHRAE 90.1* or LEC). The lighting load schedules, as a fraction of peak lighting loads, in Figure 3-13 are representative of typical residential occupant activity where the greatest loads are in the late evening between 7:00 PM and 11:00 PM. There is also a spike in lighting loads in the morning between 7:00 AM and 10:00 AM. The lighting loads also vary slightly based on the day of the week.

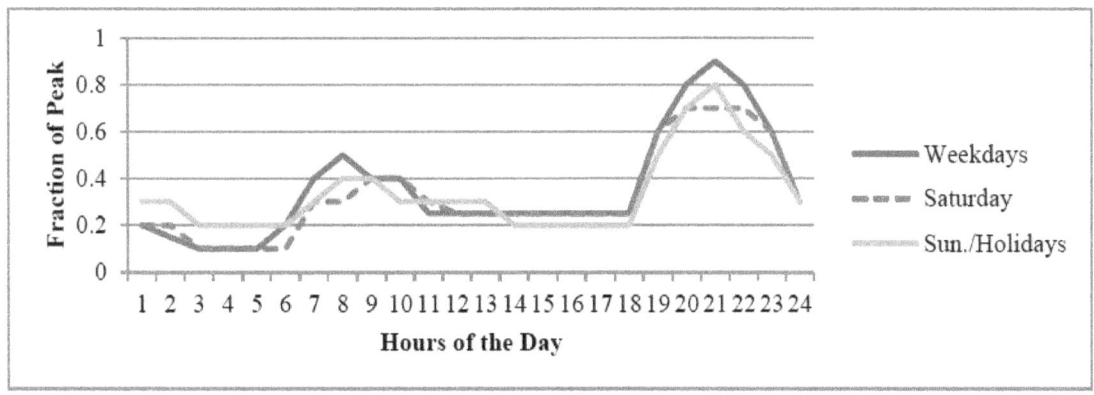

Figure 3-13 DORMI04 Lighting Schedule

The peak electrical equipment load is 8331 W, or 2.69 W/m^2 (0.25 W/ft^2). Similar to lighting loads, the electrical load schedule in Figure 3-14 is highly correlated with occupant activity.

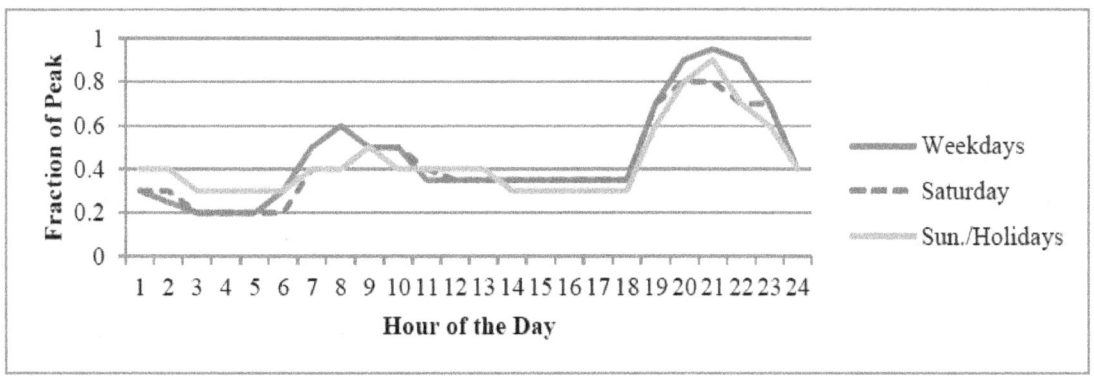

Figure 3-14 DORMI04 Electrical Load Schedule

3.3.4 6-Story Dormitory

The 6-story dormitory has steel-framed walls, insulation entirely above the roof deck, operable windows, and a window-to-wall ratio of 20 %. The detailed assumptions used in the energy simulations are described below.

Building Envelope

The energy efficiency characteristics of the building envelope are determined by the building's location and the edition of *ASHRAE 90.1*. The window characteristics (U-factor, SHGC, and VT) are based on the *ASHRAE 90.1* requirements for operable windows for 10.1 % to 20.0 % glazing.

The wall and roof efficiency characteristics are based on the *ASHRAE 90.1* requirements for residential buildings with above grade, steel-framed wall construction and insulation entirely above the roof deck.

Heating, Ventilation, and Air Conditioning

There are four main aspects to the heating, ventilation, and air conditioning of a building: equipment, operating conditions, air infiltration, and mechanical ventilation. The HVAC equipment is a packaged electric air-cooled chiller and natural gas-fired hot water boiler. Each building type that falls into the "Lodging" CBECS category has the same constant heating and cooling setpoint temperatures shown in Figure 3-15: 21 °C (70 °F) for heating and 24 °C (76 °F) for cooling.

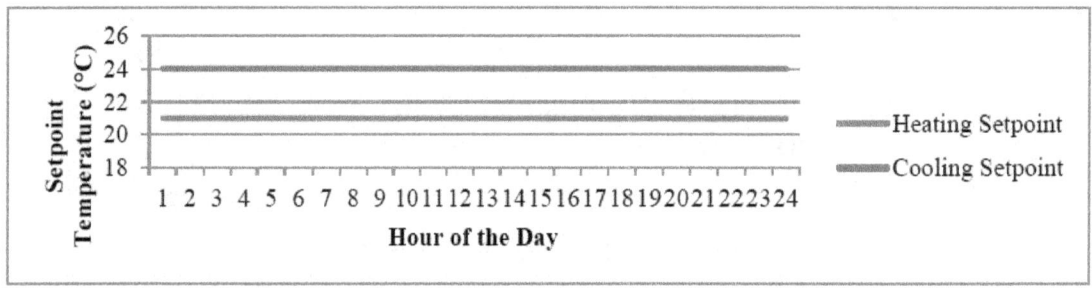

Figure 3-15 DORMI06 Setpoint Temperature Schedules

Air infiltration and mechanical ventilation are assumed to be constant across all editions of *ASHRAE 90.1*, with an infiltration rate of 0.401 m^3/s per floor (0.30 ACH) and minimum mechanical ventilation of 0.537 m^3/s per floor (0.40 ACH).

Occupancy

The peak occupancy for the 6-story dormitory is assumed to be 342 people or 1 person per 23.2 m^2 (250 ft^2). The schedule in Figure 3-16 shows that the greatest occupancy occurs over the nighttime while the lowest occupancy is during the middle of the day.

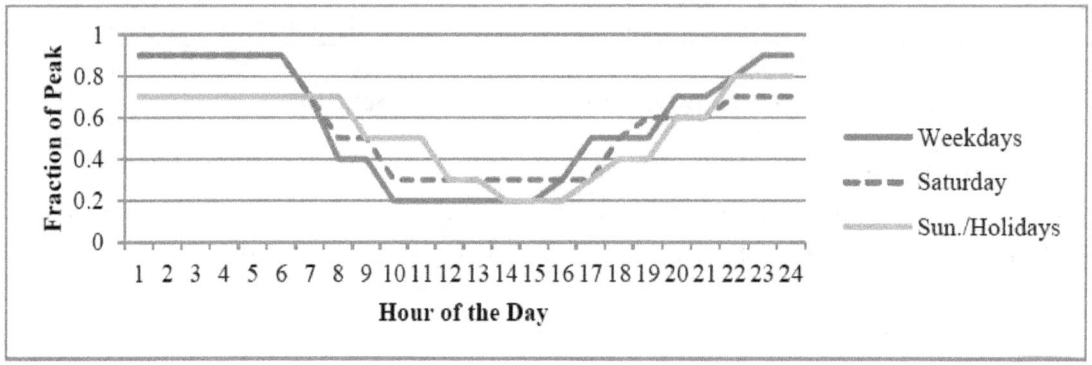

Figure 3-16 DORMI06 Occupancy Schedule

The energy simulation assumes between 8.6 W/m² (0.8 W/ft²) and 18.3 W/m² (1.7 W/ft²) of lighting density depending on the building design (e.g., edition of *ASHRAE 90.1* or LEC). The lighting load schedules, as a fraction of peak lighting loads, in Figure 3-17 are representative of typical residential occupant activity where the greatest loads are in the late evening between 7:00 PM and 11:00 PM. There is also a spike in lighting loads in the morning between 7:00 AM and 10:00 AM. The lighting loads also vary slightly based on the day of the week.

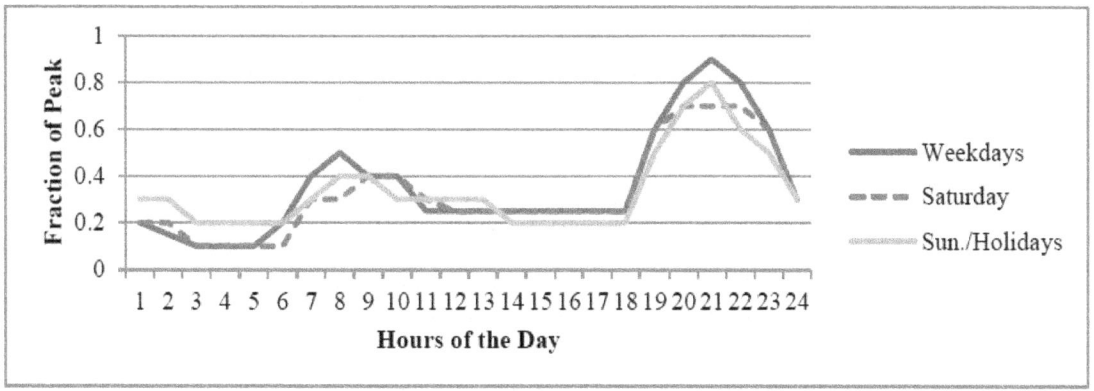

Figure 3-17 DORMI06 Lighting Schedule

The peak electrical equipment load is 21 243 W, or 2.69 W/m² (0.25 W/ft²). Similar to lighting loads, the electrical load schedule in Figure 3-18 is highly correlated with occupant activity.

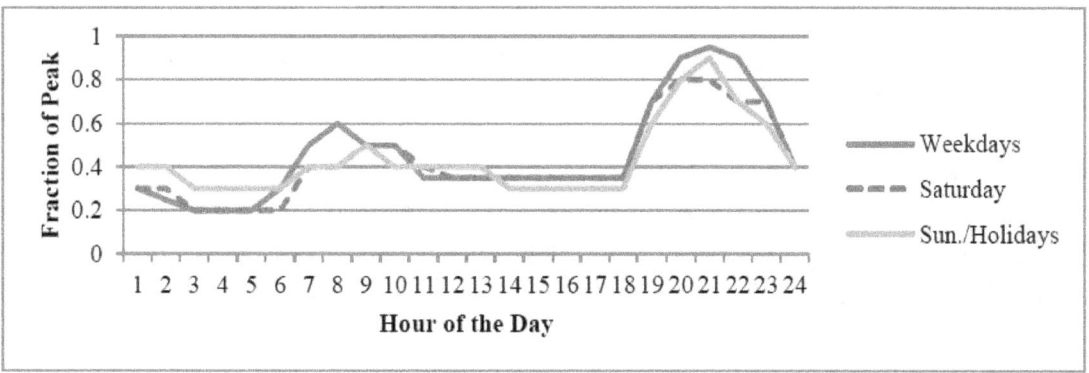

Figure 3-18 DORMI06 Electrical Load Schedule

3.3.5 15-Story Hotel

The 15-story hotel has glass and metal curtain walls with steel framing, insulation entirely above the roof deck, and a window-to-wall ratio of 100 %. The detailed assumptions used in the energy simulations are described below.

Building Envelope

The energy efficiency characteristics of the building envelope are determined by the building's location and the edition of *ASHRAE 90.1*. The window characteristics (U-factor, SHGC, and VT) are based on the *ASHRAE 90.1* requirements for fixed windows for 40.1 % to 50.0 % glazing.

The wall and roof efficiency characteristics are based on the *ASHRAE 90.1* requirements for residential buildings with above grade, steel-framed wall construction and insulation entirely above the roof deck.

Heating, Ventilation, and Air Conditioning

There are four main aspects to the heating, ventilation, and air conditioning of a building: equipment, operating conditions, air infiltration, and mechanical ventilation. The HVAC equipment is a packaged electric water-cooled chiller and natural gas-fired hot water boiler. Each building type that falls into the "Lodging" CBECS category has the same constant heating and cooling setpoint temperatures shown in Figure 3-19: 21 °C (70 °F) for heating and 24 °C (76 °F) for cooling.

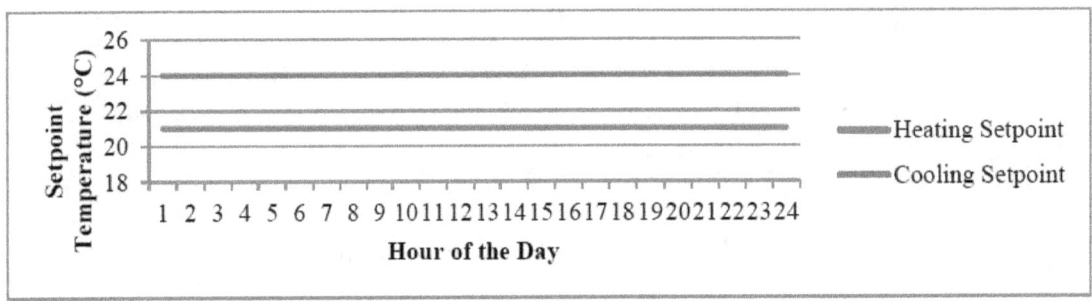

Figure 3-19 HOTEL15 Setpoint Temperature Schedules

Air infiltration and mechanical ventilation are assumed to be constant across all editions of *ASHRAE 90.1*, with an infiltration rate of 0.708 m³/s per floor (0.30 ACH) and minimum mechanical ventilation of 1.136 m³/s per floor (0.48 ACH).

Occupancy, Lighting, and Electrical Loads

The peak occupancy for the 15-story hotel is assumed to be 1800 people or 1 person per 23.2 m² (250 ft²). The schedule in Figure 3-20 shows that the greatest occupancy occurs over the nighttime while the lowest occupancy is during the middle of the day.

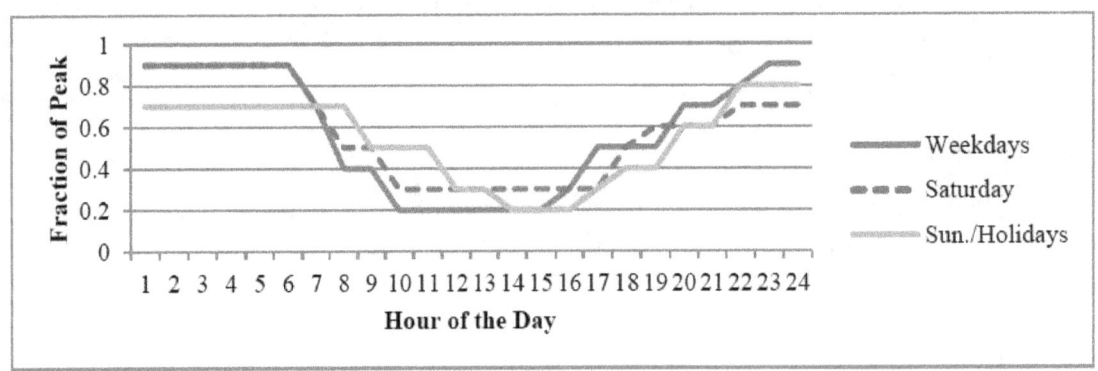

Figure 3-20 HOTEL15 Occupancy Schedule

The energy simulation assumes between 8.6 W/m² (0.8 W/ft²) and 18.3 W/m² (1.7 W/ft²) of lighting density depending on the building design (e.g., edition of *ASHRAE 90.1* or LEC). The lighting load schedules, as a fraction of peak lighting loads, in Figure 3-21 are representative of typical residential occupant activity where the greatest loads are in the late evening between 7:00 PM and 11:00 PM. There is also a spike in lighting loads in the morning between 7:00 AM and 10:00 AM. The lighting loads also vary slightly based on the day of the week.

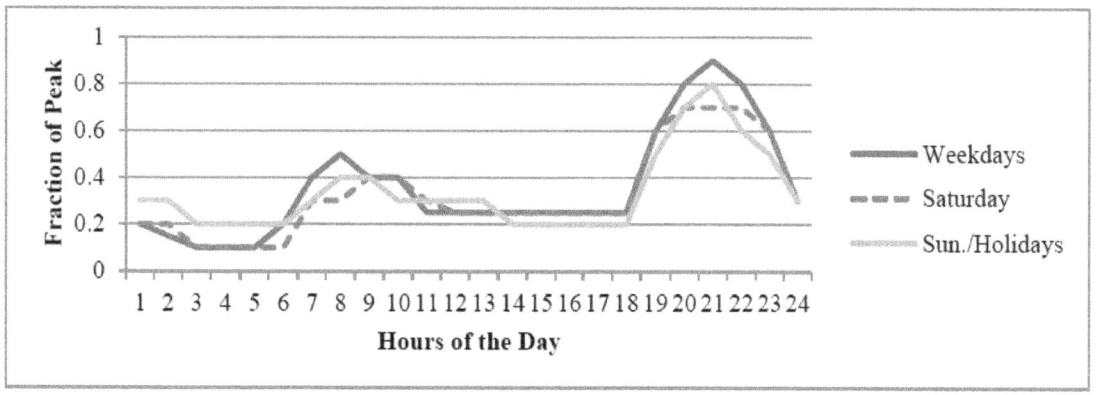

Figure 3-21 HOTEL15 Lighting Schedule

The peak electrical equipment load is 112 462 W, or 2.69 W/m² (0.25 W/ft²). Similar to lighting loads, the electrical load schedule in Figure 3-22 is highly correlated with occupant activity.

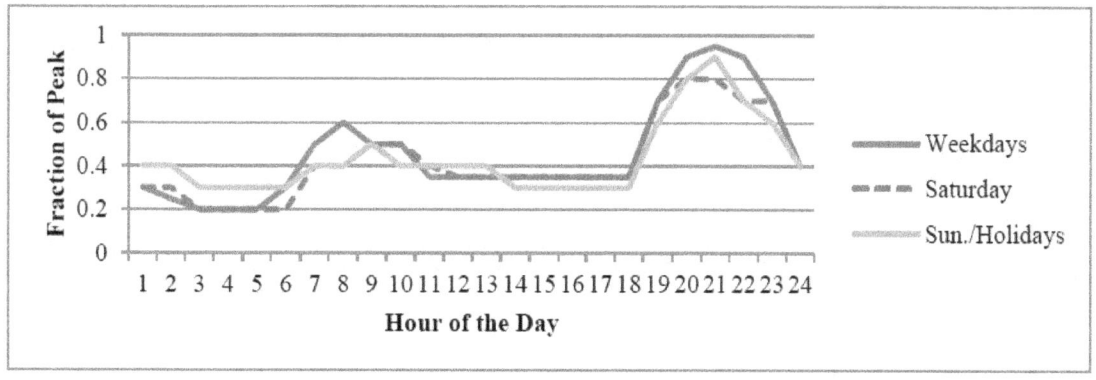

Figure 3-22 HOTEL15 Electrical Load Schedule

3.3.6 2-Story High School

The 2-story high school has mass walls, insulation entirely above the roof deck, operable windows, and a window-to-wall ratio of 25 %. The detailed assumptions used in the energy simulations are described below.

Building Envelope

The energy efficiency characteristics of the building envelope are determined by the building's location and the edition of *ASHRAE 90.1*. The window characteristics (U-factor, SHGC, and VT) are based on the *ASHRAE 90.1* requirements for operable windows for 20.1 % to 30.0 % glazing.

The wall and roof efficiency characteristics are based on the *ASHRAE 90.1* requirements for nonresidential buildings with above grade, mass wall construction and insulation entirely above the roof deck.

Heating, Ventilation, and Air Conditioning

There are four main aspects to the heating, ventilation, and air conditioning of a building: equipment, operating conditions, air infiltration, and mechanical ventilation. The HVAC equipment is a water-cooled chiller and a natural gas-fired hot water boiler. Each building type that falls into the "Education" CBECS category has the same heating and cooling setpoint temperature schedules. Figure 3-23 shows that the heating setpoint temperature is a constant 16 °C (61 °F) for weekends while it is 21 °C (70 °F) during the daytime and 16 °C (61 °F) during the nighttime on weekdays. Similarly, Figure 3-24 shows that the cooling setpoint temperature is a constant 31 °C (88 °F) for weekends while it is 25°C (77 °F) during the daytime and 31°C (88 °F) during the nighttime weekdays. These setpoints correlate with the building occupancy schedule.

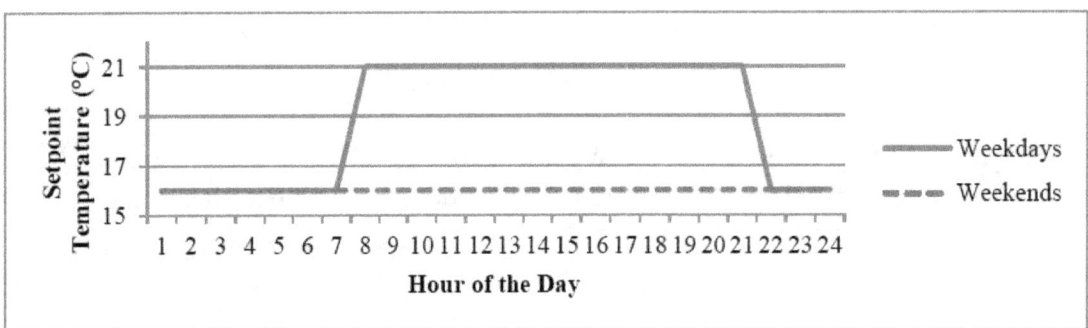

Figure 3-23 HIGHS02 Heating Setpoint Temperature Schedule

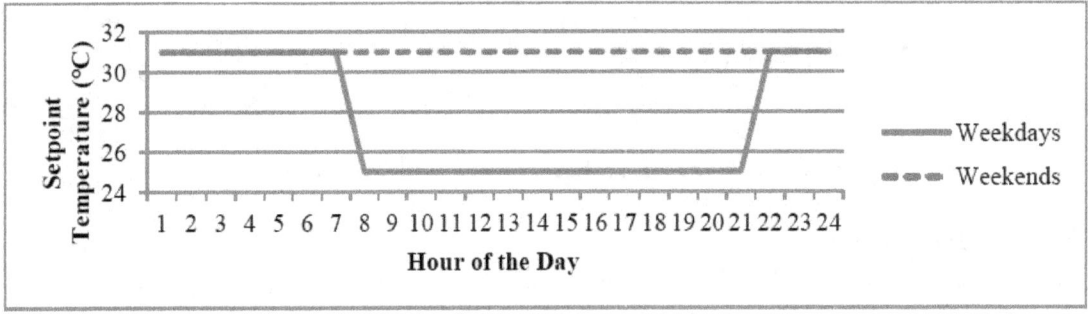

Figure 3-24 HIGHS02 Cooling Setpoint Temperature Schedule

Air infiltration and mechanical ventilation are assumed to be constant across all editions of *ASHRAE 90.1*, with an infiltration rate of 2.301 m³/s per floor (0.30 ACH) and minimum mechanical ventilation of 7.971 m³/s per floor (1.04 ACH).

Occupancy, Lighting, and Electrical Loads

The peak occupancy for the 2-story high school is assumed to be 1740 people or 1 person per 7.0 m^2 (75 ft^2). The schedule in Figure 3-25 shows that the greatest occupancy occurs on the weekdays during the school year. The high school is assumed to not be used during the nighttime or on weekends year-round. On weekdays during the summer, the occupancy is much lower than during the school year.

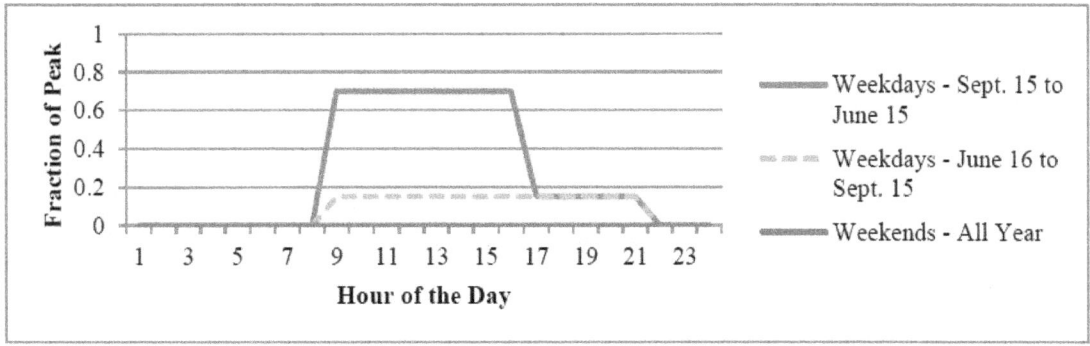

Figure 3-25 HIGHS02 Occupancy Schedule

The energy simulation assumes between 10.8 W/m^2 (1.0 W/ft^2) and 16.1 W/m^2 (1.5 W/ft^2) of lighting density depending on the building design (e.g., edition of *ASHRAE 90.1* or LEC). The lighting load schedules, as a fraction of peak lighting loads, in Figure 3-26 are representative of typical school occupant activity. The loads are greatest during daytime hours on weekdays of the school year. Daytime loads are lower during the summer while school is out. There is no lighting use during the nighttime or on weekends year-round.

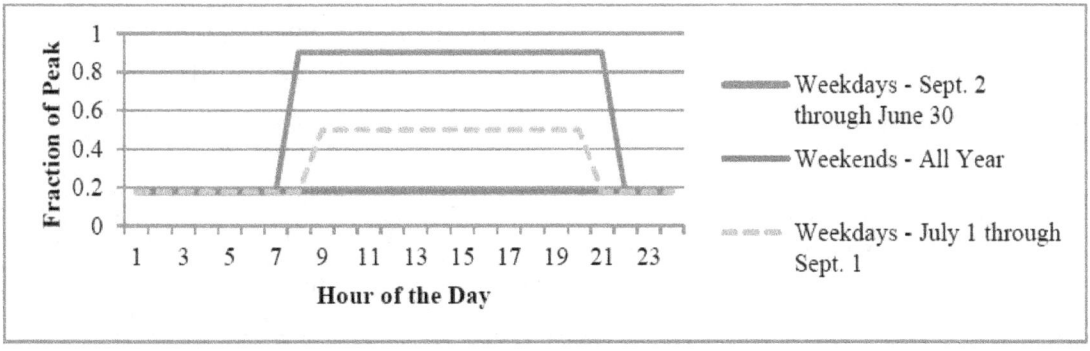

Figure 3-26 HIGHS02 Lighting Schedule

The peak electrical equipment load is 64 978 W, or 5.38 W/m^2 (0.5 W/ft^2). The electrical load schedule in Figure 3-27 is highly correlated with the times of the year that children and teachers are at the school (9 AM to 5 PM).

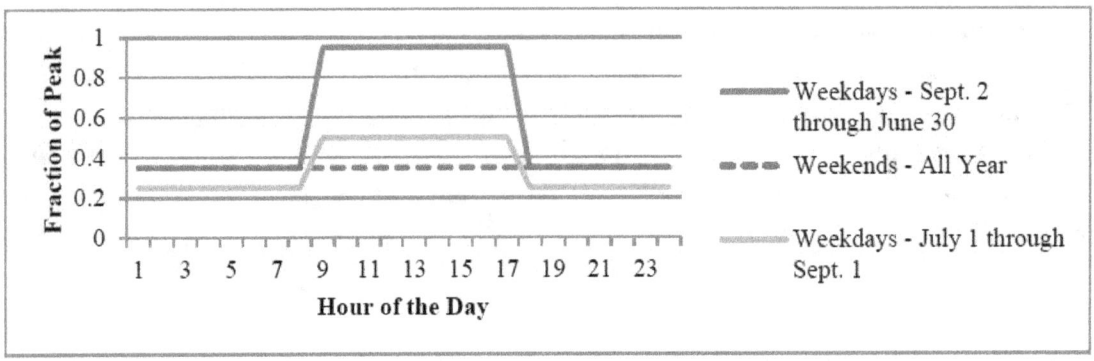

Figure 3-27 HIGHS02 Electrical Load Schedule

3.3.7 3-Story Office Building

The 3-story office building has mass walls, insulation entirely above the roof deck, operable windows, and a window-to-wall ratio of 20 %. The detailed assumptions used in the energy simulations are described below.

Building Envelope

The energy efficiency characteristics of the building envelope are determined by the building's location and the edition of *ASHRAE 90.1*. The window characteristics (U-factor, SHGC, and VT) are based on the *ASHRAE 90.1* requirements for operable windows for 10.1 % to 20.0 % glazing. The wall and roof efficiency characteristics are based on the *ASHRAE 90.1* requirements for nonresidential buildings with above grade, mass wall construction and insulation entirely above the roof deck.

Heating, Ventilation, and Air Conditioning

There are four main aspects to the heating, ventilation, and air conditioning of a building: equipment, operating conditions, air infiltration, and mechanical ventilation. The HVAC equipment is an air-cooled electric chiller and a natural gas-fired hot water boiler. Each building type that falls into the "Office" CBECS category has the same heating and cooling setpoint temperature schedules. Figure 3-28 shows that the heating setpoint temperature varies by day of the week. The setpoint is a constant 15.6 °C (60 °F) for Sundays and holidays while it is 21 °C (70 °F) from 7 AM to 5 PM and 15.6 °C (60 °F) for the rest of the day on Saturdays. Weekdays have a similar schedule to Saturdays, with a setpoint of 21 °C (70 °F) from 6 AM to 7 PM and 15.6 °C (60 °F) for all other hours. Similarly, Figure 3-29 shows that the cooling setpoint temperature is a constant 30 °C (86 °F) on Sundays and holidays while it is 24 °C (75 °F) from 7 AM to 6 PM and 30 °C (86 °F) the remainder of the day on Saturdays. Weekdays have a setpoint of 24 °C (75 °F) from 7 AM to 10 PM and 30 °C (86 °F) for all other hours. These setpoints correlate with the building occupancy schedule.

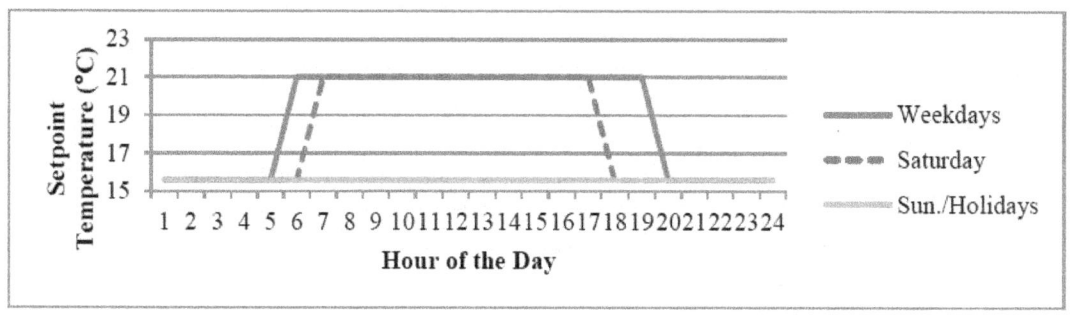

Figure 3-28 OFFIC03 Heating Setpoint Temperature Schedule

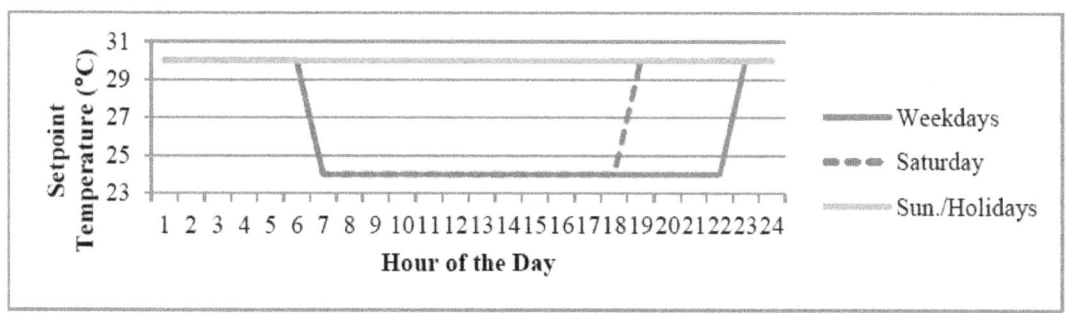

Figure 3-29 OFFIC03 Cooling Setpoint Temperature Schedule

Air infiltration and mechanical ventilation are assumed to be constant across all editions of *ASHRAE 90.1*, with an infiltration rate of 0.189 m³/s per floor (0.30 ACH) and minimum mechanical ventilation of 0.246 m³/s per floor (0.39 ACH).

Occupancy, Lighting, and Electrical Loads

The peak occupancy for the 3-story office building is assumed to be 72 people or 1 person per 25.5 m² (275 ft²). The schedule in Figure 3-30 shows that occupancy represents typical office activity, where the majority of people are in the building during typical "office hours" (9 AM to 5 PM) on the weekdays with a drop in occupancy over the lunch hour. There is a relatively small amount of occupant activity on Saturdays during typical "office hours." There is no occupancy on Sundays or holidays.

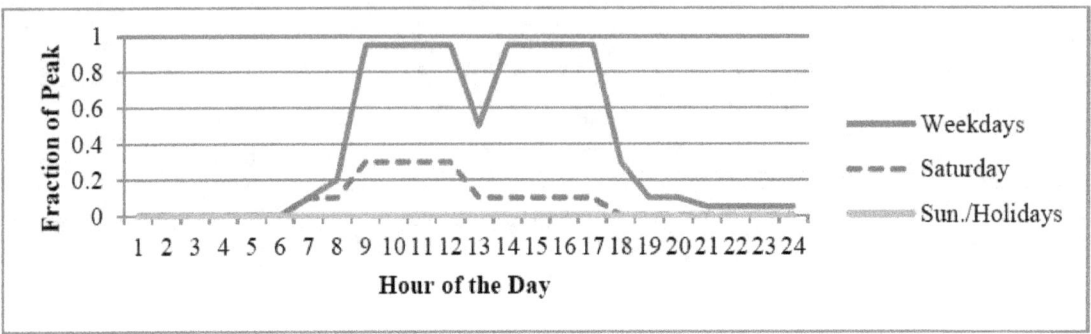

Figure 3-30 OFFIC03 Occupancy Schedule

The energy simulation assumes between 8.6 W/m^2 (0.8W/ft^2) and 14.0 W/m^2 (1.3 W/ft^2) of lighting density depending on the building design (e.g., edition of *ASHRAE 90.1* or LEC). The lighting load schedules, as a fraction of peak lighting loads, in Figure 3-31 are representative of typical office occupant activity. The loads are greatest during "working" hours on weekdays. Daytime loads are lower on Saturdays. There is no lighting use during the nighttime or on Sundays or holidays.

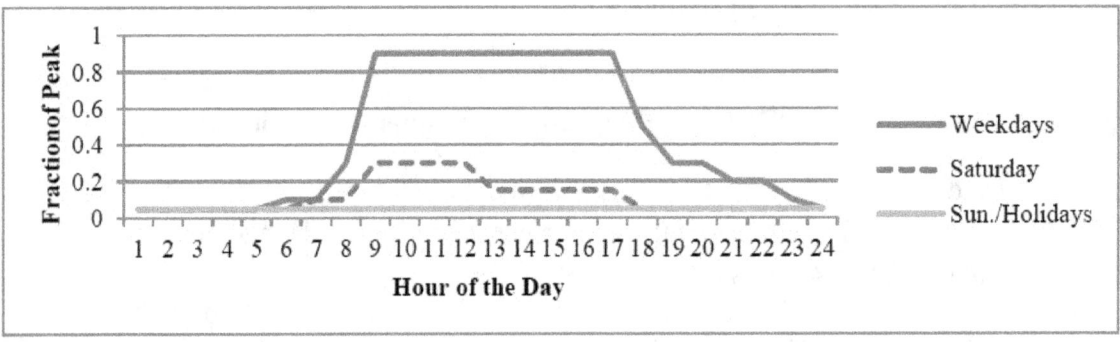

Figure 3-31 OFFIC03 Lighting Schedule

The peak electrical equipment load is 14 996 W, or 8.07 W/m^2 (0.75 W/ft^2). The electrical load schedule in Figure 3-32 is highly correlated with the occupancy schedule with the greatest electrical loads between 9 AM to 5 PM on weekdays.

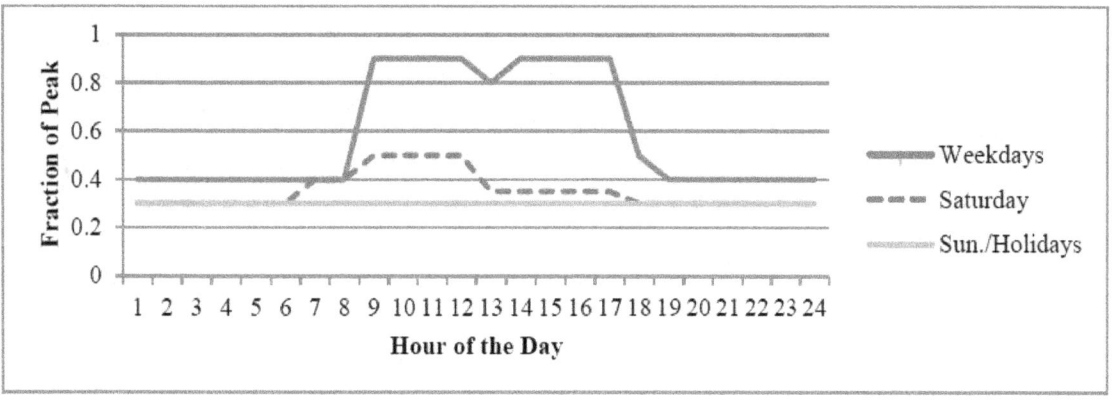

Figure 3-32 OFFIC03 Electrical Load Schedule

3.3.8 8-Story Office Building

The 8-story office building has mass walls, insulation entirely above the roof deck, operable windows, and a window-to-wall ratio of 20 %. The detailed assumptions used in the energy simulations are described below.

Building Envelope

The energy efficiency characteristics of the building envelope are determined by the building's location and the edition of *ASHRAE 90.1*. The window characteristics (U-factor, SHGC, and VT) are based on the *ASHRAE 90.1* requirements for operable windows for 10.1 % to 20.0 % glazing. The wall and roof efficiency characteristics are based on the *ASHRAE 90.1* requirements for nonresidential buildings with above grade, mass wall construction and insulation entirely above the roof deck.

Heating, Ventilation, and Air Conditioning

There are four main aspects to the heating, ventilation, and air conditioning of a building: equipment, operating conditions, air infiltration, and mechanical ventilation. The HVAC equipment is a rooftop packaged air conditioner and a natural gas-fired furnace. Each building type that falls into the "Office" CBECS category has the same heating and cooling setpoint temperature schedules. Figure 3-33 shows that the heating setpoint temperature varies by day of the week. The setpoint is a constant 15.6 °C (60 °F) for Sundays and holidays while it is 21 °C (70 °F) from 7 AM to 5 PM and 15.6 °C (60 °F) for the rest of the day on Saturdays. Weekdays have a similar schedule to Saturdays, with a setpoint of 21 °C (70 °F) from 6 AM to 7 PM and 15.6 °C (60 °F) for all other hours. Similarly, Figure 3-34 shows that the cooling setpoint temperature is a constant 30C on Sundays and holidays while it is 24 °C (75 °F) from 7 AM to 6 PM and 30 °C (86 °F) the remainder of the day on Saturdays. Weekdays have a setpoint of 24 °C (75 °F) from 7 AM to 10 PM and 30 °C (86 °F) for all other hours. These setpoints correlate with the building occupancy schedule.

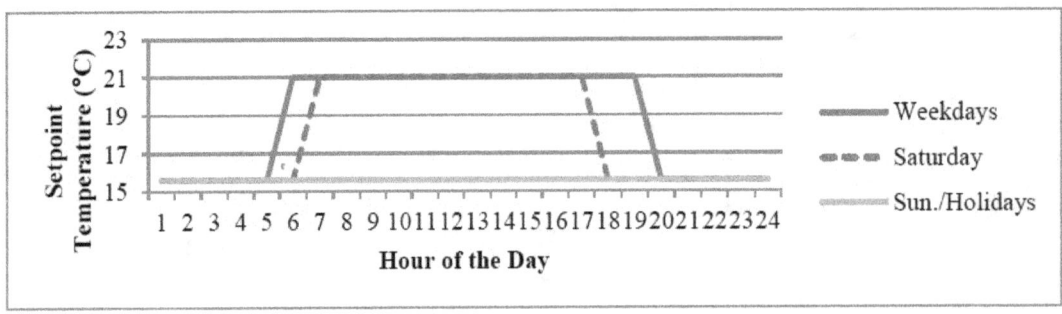

Figure 3-33 OFFIC08 Heating Setpoint Temperature Schedule

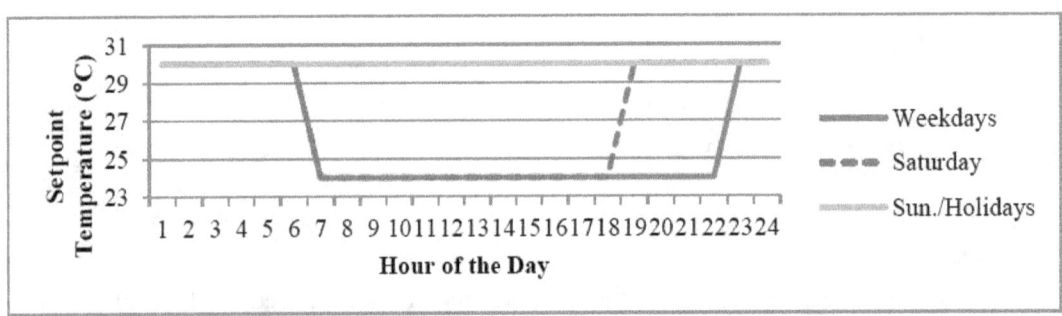

Figure 3-34 OFFIC08 Cooling Setpoint Temperature Schedule

Air Infiltration and Mechanical Ventilation

Air infiltration and mechanical ventilation are assumed to be constant across all editions of *ASHRAE 90.1*, with an infiltration rate of 0.283 m³/s per floor (0.30 ACH) and minimum mechanical ventilation of 0.370 m³/s per floor (0.39 ACH).

Occupancy, Lighting, and Electrical Loads

The peak occupancy for the 8-story office building is assumed to be 288 people or 1 person per 25.5 m² (275 ft²). The schedule in Figure 3-35 shows that occupancy represents typical office activity, where the majority of people are in the building during typical "office hours" (9 AM to 5 PM) on the weekdays with a drop in occupancy over the lunch hour. There is a relatively small amount of occupant activity on Saturdays during typical "office hours." There is no occupancy on Sundays or holidays.

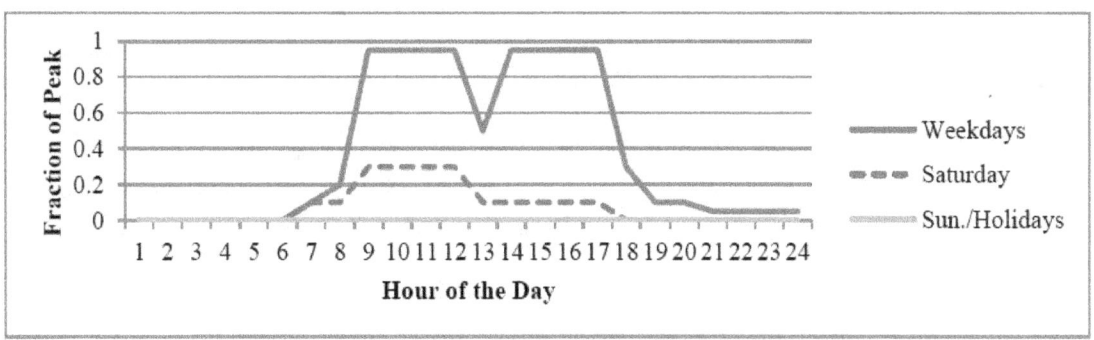

Figure 3-35 OFFIC08 Occupancy Schedule

The energy simulation assumes between 8.6 W/m^2 (0.8W/ft^2) and 14.0 W/m^2 (1.3 W/ft^2) of lighting density depending on the building design (e.g., edition of *ASHRAE 90.1* or LEC). The lighting load schedules, as a fraction of peak lighting loads, in Figure 3-36 are representative of typical office occupant activity. The loads are greatest during "working" hours on weekdays. Daytime loads are lower on Saturdays. There is no lighting use during the nighttime or on Sundays or holidays.

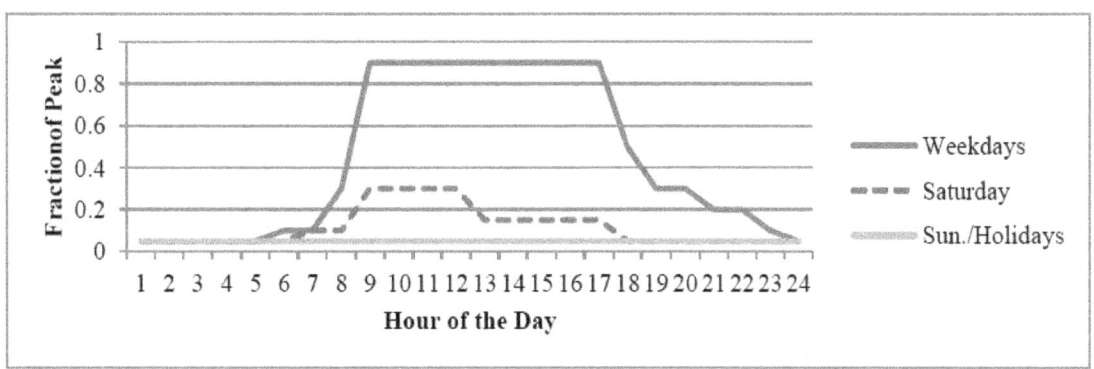

Figure 3-36 OFFIC08 Lighting Schedule

The peak electrical equipment load is 59 978 W, or 8.07 W/m^2 (0.75 W/ft^2). The electrical load schedule in Figure 3-37 is highly correlated with the occupancy schedule with the greatest electrical loads between 9 AM to 5 PM on weekdays.

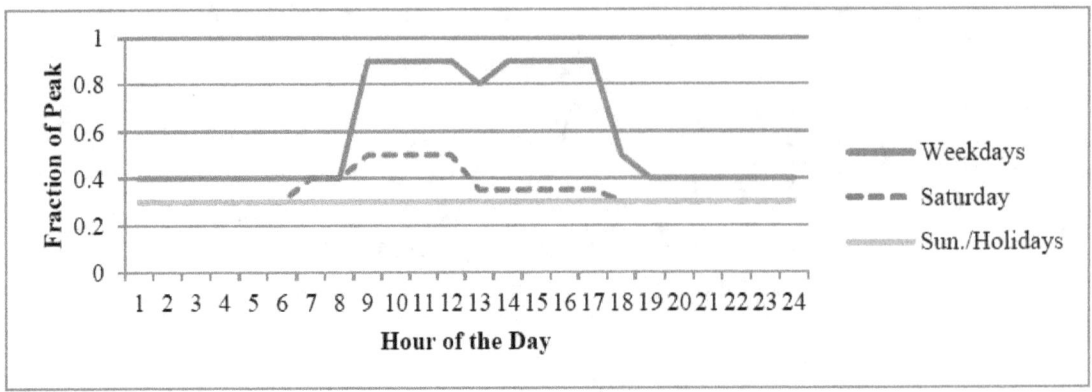

Figure 3-37 OFFIC08 Electrical Load Schedule

3.3.9 16-Story Office Building

The 16-story office building has glass and metal curtain walls with steel framing, insulation entirely above the roof deck, and a window-to-wall ratio of 100 %. The detailed assumptions used in the energy simulations are described below.

Building Envelope

The energy efficiency characteristics of the building envelope are determined by the building's location and the edition of *ASHRAE 90.1*. The window characteristics (U-factor, SHGC, and VT) are based on the *ASHRAE 90.1* requirements for operable windows for 40.1 % to 50.0 % glazing. The wall and roof efficiency characteristics are based on the *ASHRAE 90.1* requirements for nonresidential buildings with above grade, steel-framed wall construction and insulation entirely above the roof deck.

Heating, Ventilation, and Air Conditioning

There are four main aspects to the heating, ventilation, and air conditioning of a building: equipment, operating conditions, air infiltration, and mechanical ventilation. The HVAC equipment is a water-cooled electric chiller and a natural gas-fired hot-water boiler. Each building type that falls into the "Office" CBECS category has the same heating and cooling setpoint temperature schedules. Figure 3-38 shows that the heating setpoint temperature varies by day of the week. The setpoint is a constant 15.6 °C (60 °F) for Sundays and holidays while it is 21 °C (70 °F) from 7 AM to 5 PM and 15.6 °C (60 °F) for the rest of the day on Saturdays. Weekdays have a similar schedule to Saturdays, with a setpoint of 21 °C (70 °F) from 6 AM to 7 PM and 15.6 °C (60 °F) for all other hours. Similarly, Figure 3-39 shows that the cooling setpoint temperature is a constant 30 °C (86 °F) on Sundays and holidays while it is 24 °C (75 °F) from 7 AM to 6 PM and 30 °C (86 °F) the remainder of the day on Saturdays. Weekdays have a setpoint of 24 °C (75 °F) from 7 AM to 10 PM and 30 °C (86 °F) for all other hours. These setpoints correlate with the building occupancy schedule.

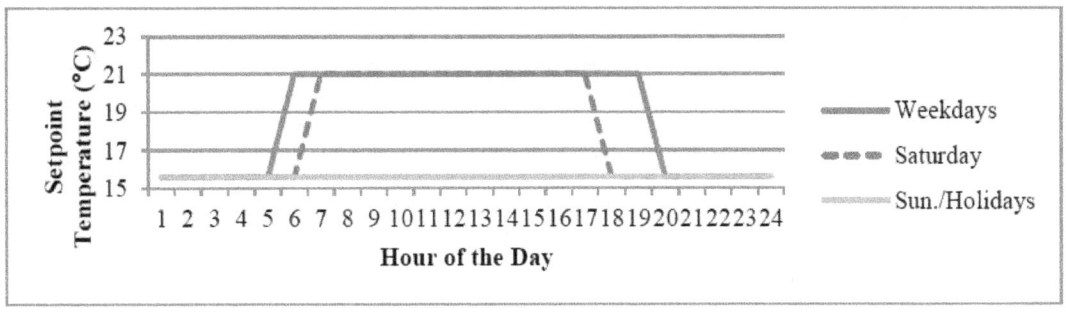

Figure 3-38 OFFIC16 Heating Setpoint Temperature Schedule

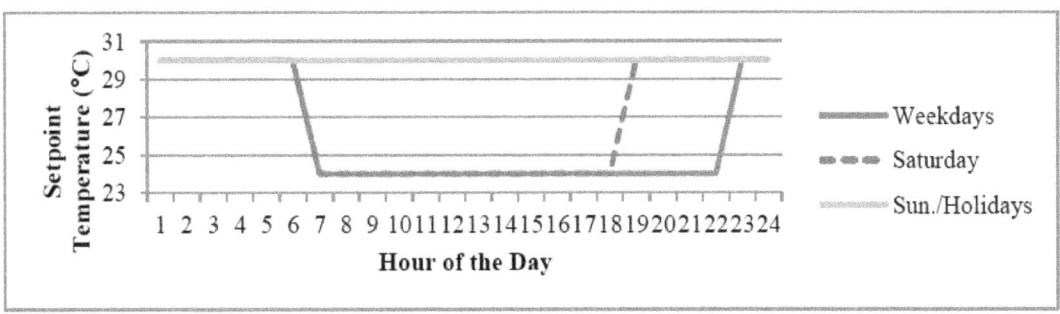

Figure 3-39 OFFIC16 Cooling Setpoint Temperature Schedule

Air infiltration and mechanical ventilation are assumed to be constant across all editions of *ASHRAE 90.1*, with an infiltration rate of 0.384 m³/s per floor (0.30 ACH) and minimum mechanical ventilation of 0.600 m³/s per floor (0.47 ACH).

Occupancy, Lighting, and Electrical Loads

The peak occupancy for the 16-story office building is assumed to be 944 people or 1 person per 25.5 m² (275 ft²). The schedule in Figure 3-40 shows that occupancy represents typical office activity, where the majority of people are in the building during typical "office hours" (9 AM to 5 PM) on the weekdays with a drop in occupancy over the lunch hour. There is a relatively small amount of occupant activity on Saturdays during typical "office hours." There is no occupancy on Sundays or holidays.

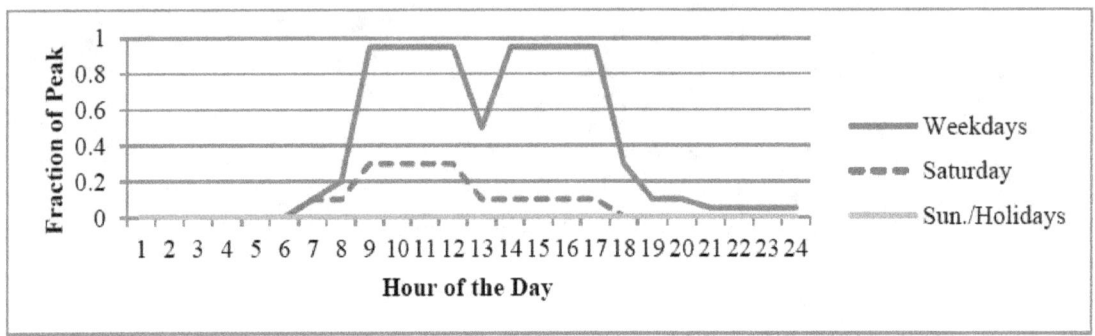

Figure 3-40 OFFIC16 Occupancy Schedule

The energy simulation assumes between 8.6 W/m^2 (0.8W/ft^2) and 14.0 W/m^2 (1.3 W/ft^2) of lighting density depending on the building design (e.g., edition of *ASHRAE 90.1* or LEC). The lighting load schedules, as a fraction of peak lighting loads, in Figure 3-41 are representative of typical office occupant activity. The loads are greatest during "working" hours on weekdays. Daytime loads are lower on Saturdays. There is no lighting use during the nighttime or on Sundays or holidays.

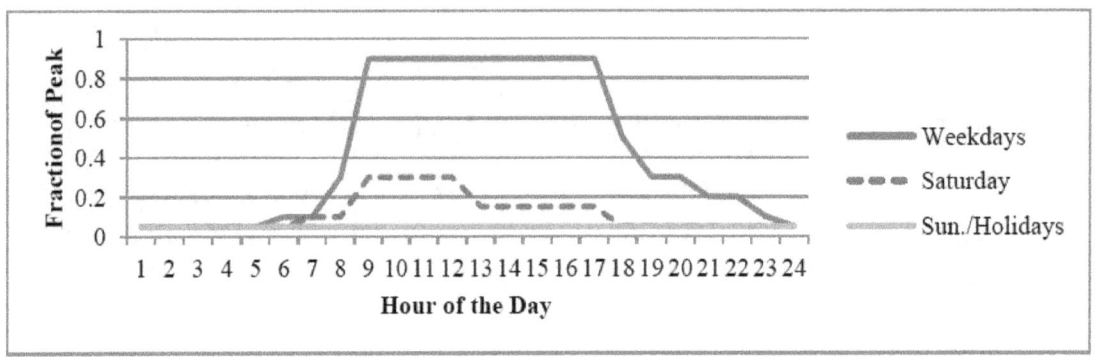

Figure 3-41 OFFIC16 Lighting Schedule

The peak electrical equipment load is 194 934 W, or 8.07 W/m^2 (0.75 W/ft^2). The electrical load schedule in Figure 3-42 is highly correlated with the occupancy schedule with the greatest electrical loads between 9 AM to 5 PM on weekdays.

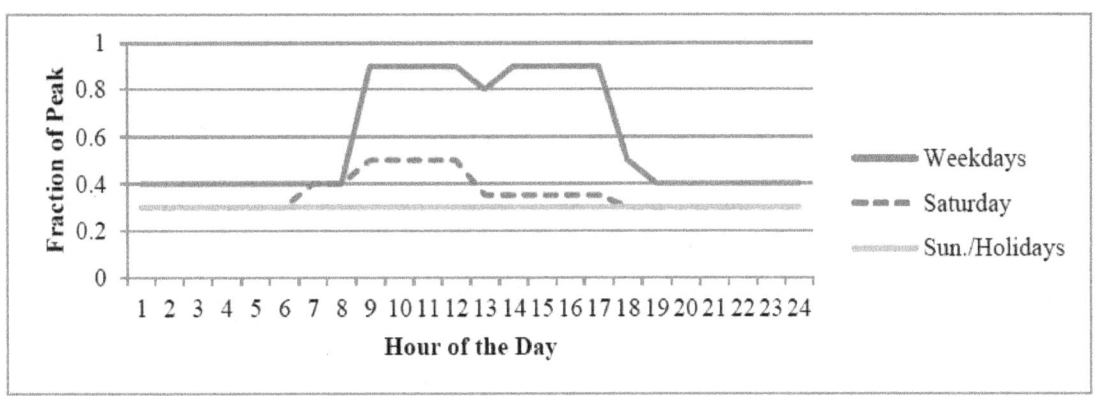

Figure 3-42 OFFIC16 Electrical Load Schedule

3.3.10 1-Story Retail Store

The 1-story retail store has mass walls, insulation entirely above the roof deck, fixed windows, and a window-to-wall ratio of 10 %. The detailed assumptions used in the energy simulations are described below.

Building Envelope

The energy efficiency characteristics of the building envelope are determined by the building's location and the edition of *ASHRAE 90.1*. The window characteristics (U-factor, SHGC, and VT) are based on the *ASHRAE 90.1* requirements for operable windows for 0.0 % to 10.0 % glazing. The wall and roof efficiency characteristics are based on the *ASHRAE 90.1* requirements for nonresidential buildings with above grade, mass wall construction and insulation entirely above the roof deck.

Heating, Ventilation, and Air Conditioning

There are four main aspects to the heating, ventilation, and air conditioning of a building: equipment, operating conditions, air infiltration, and mechanical ventilation. The HVAC equipment is a rooftop packaged electric air conditioner and a natural gas-fired furnace. Figure 3-43 show that the heating setpoint temperature for the retail store varies slightly by day of the week. For all days, the setpoint is 21 °C (70 °F) while the store is open and 15.6 °C (60 °F) when the store is closed. The store is open from 7 AM to 10 PM on weekdays and 7 AM to 11 PM on Saturdays. The store hour on Sundays and holidays are 9 AM to 8 PM. These setpoints correlate with the building occupancy schedule. Figure 3-44 shows a nearly identical pattern for the cooling setpoint temperature. The setpoint is 24 °C (75 °F) for 7 AM to 9 PM on weekdays, 7 AM to 10 PM on Saturdays, and 9 AM to 7 PM on Sundays and holidays. The setpoint is 30 °C (86 °F) while the store is closed.

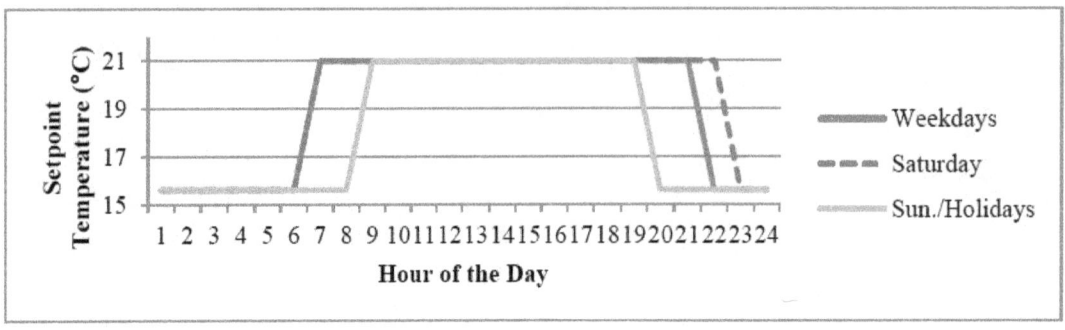

Figure 3-43 RETAIL1 Heating Setpoint Temperature Schedule

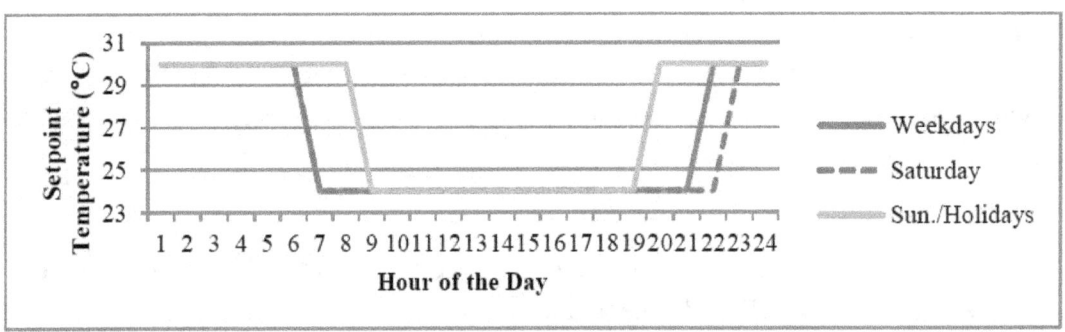

Figure 3-44 RETAIL1 Cooling Setpoint Temperature Schedule

Air infiltration and mechanical ventilation are assumed to be constant across all editions of *ASHRAE 90.1*, with an infiltration rate of 0.378 m³/s per floor (0.43 ACH) and minimum mechanical ventilation of 0.547 m³/s per floor (0.62 ACH).

Occupancy, Lighting, and Electrical Loads

The peak occupancy for the 1-story retail store is assumed to be 27 people or 1 person per 25.5 m² (300 ft²). The schedule in Figure 3-45 shows that occupancy varies significantly both within a given day and across days of the week. In general, the afternoon and early evening is the busy time over all days, which is the most common time of the day for people to shop. The afternoon is the busiest time on the weekends and holidays while the early evening has the greatest occupancy on weekdays. The greatest occupancy occurs on Saturday followed by weekdays and then Sundays and holidays.

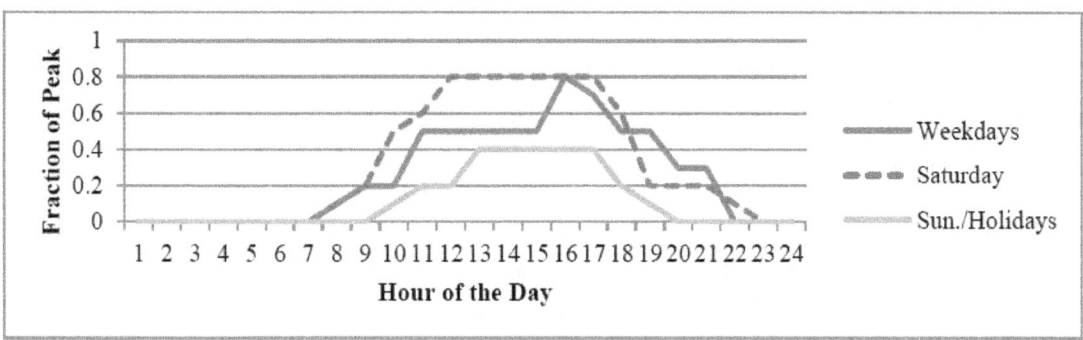

Figure 3-45 RETAIL1 Occupancy Schedule

The energy simulation assumes between 16.1 W/m^2 (1.5 W/ft^2) and 20.5 W/m^2 (1.9 W/ft^2) of lighting density depending on the building design (e.g., edition of *ASHRAE 90.1* or LEC). The lighting load schedule, as a fraction of peak lighting loads, in Figure 3-46 is highly correlated with the occupancy schedule. However, lighting is an on/off decision. So when the retail store is open, the lighting load is fairly constant and has less variability than the occupancy schedule.

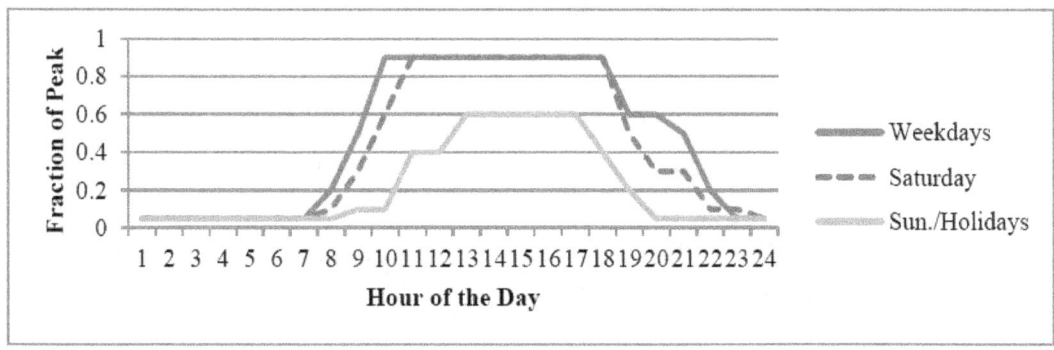

Figure 3-46 RETAIL1 Lighting Schedule

The peak electrical equipment load is 1999 W, or 2.69 W/m^2 (0.25 W/ft^2). The electrical load schedule in Figure 3-47 is highly correlated with the occupancy and lighting schedules, but has less variability across days of the week because most electrical loads are constant when a building is occupied.

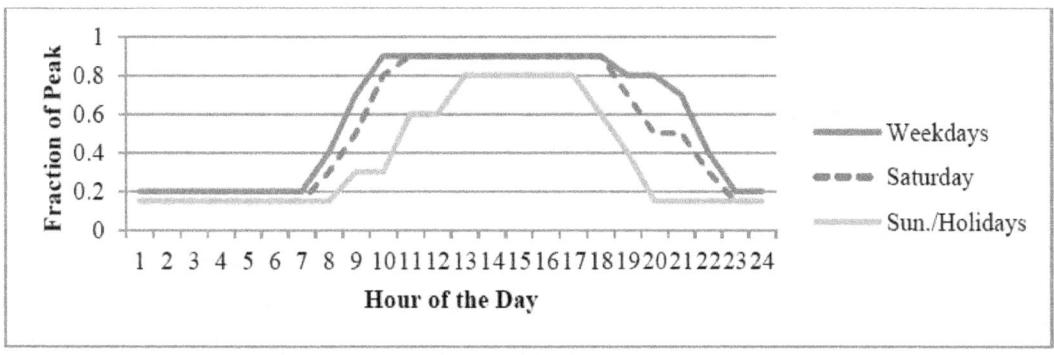

Figure 3-47 RETAIL1 Electrical Load Schedule

3.3.11 1-Story Restaurant

The 1-story restaurant has wood frame wall construction, insulation entirely above the roof deck, fixed windows, and a window-to-wall ratio of 30 %. The detailed assumptions used in the energy simulations are described below.

Building Envelope

The energy efficiency characteristics of the building envelope are determined by the building's location and the edition of *ASHRAE 90.1*. The window characteristics (U-factor, SHGC, and VT) are based on the *ASHRAE 90.1* requirements for operable windows for 20.1 % to 30.0 % glazing. The wall and roof efficiency characteristics are based on the *ASHRAE 90.1* requirements for nonresidential buildings with above grade, wood-framed wall construction and insulation entirely above the roof deck.

Heating, Ventilation, and Air Conditioning

There are four main aspects to the heating, ventilation, and air conditioning of a building: equipment, operating conditions, air infiltration, and mechanical ventilation. The HVAC equipment is a rooftop packaged electric air conditioner and a natural gas-fired furnace. Figure 3-48 shows that the heating setpoint temperature for the restaurant varies slightly by day of the week. The setpoint is 21 °C (70 °F) from midnight to 3 AM and 7 AM to midnight on weekdays, midnight to 3 AM and 9 AM to midnight on Saturdays, and midnight to 3 AM and 10 AM to midnight on Sundays and holidays. The heating setpoint is 15.6 °C (60 °F) for all other times. Figure 3-49 shows a mirror image for the cooling setpoint temperature. The setpoint is 24 °C (75 °F) from midnight to 3 AM and 7 AM to midnight on weekdays, midnight to 3 AM and 9 AM to midnight on Saturdays, and midnight to 3 AM and 10 AM to midnight on Sundays and holidays. The cooling setpoint is 30 °C (86 °F) for all other times.

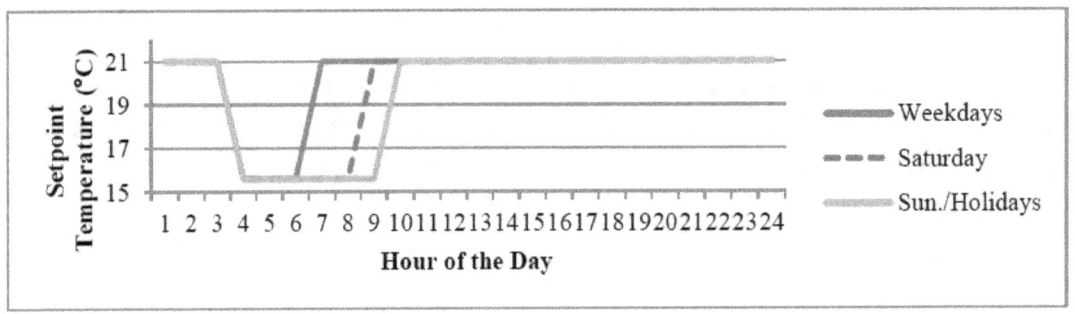

Figure 3-48 RSTRNT1 Heating Setpoint Temperature Schedule

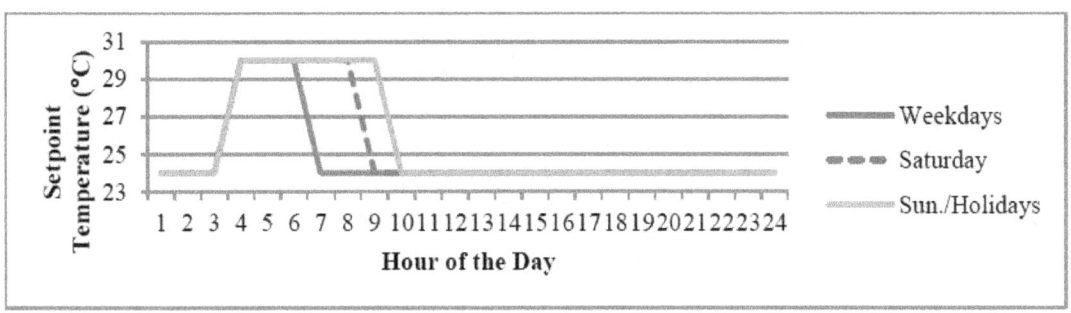

Figure 3-49 RSTRNT1 Cooling Setpoint Temperature Schedule

Air infiltration and mechanical ventilation are assumed to be constant across all editions of *ASHRAE 90.1*, with an infiltration rate of 0.275 m^3/s per floor (0.58 ACH) and minimum mechanical ventilation of 0.609 m^3/s per floor (1.29 ACH).

Occupancy, Lighting, and Electrical Loads
The peak occupancy for the 1-story restaurant is assumed to be 50 people or 1 person per 25.5 m^2 (100 ft^2). The schedule in Figure 3-50 shows that occupancy varies significantly both within a given day and across days of the week. As would be expected, lunchtime and dinnertime are the busiest times over all days.

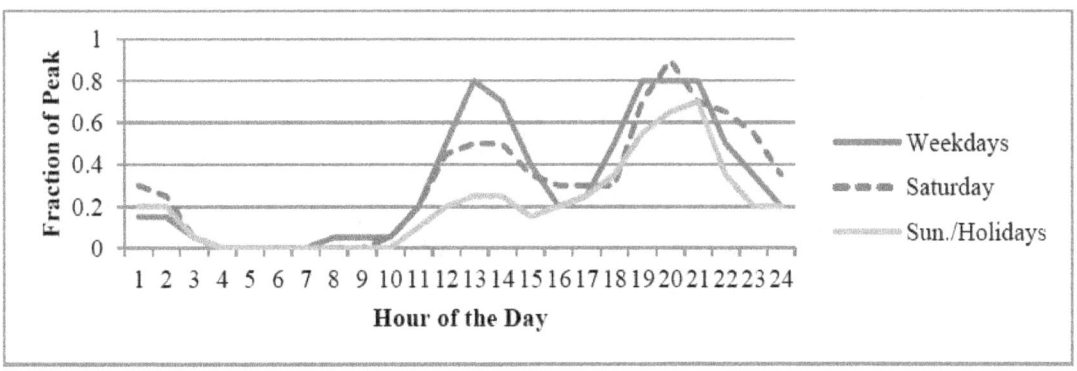

Figure 3-50 RSTRNT1 Occupancy Schedule

The energy simulation assumes between 14.0 W/m^2 (1.3 W/ft^2) and 19.4 W/m^2 (1.8 W/ft^2) of lighting density depending on the building design (e.g., edition of *ASHRAE 90.1* or LEC). The lighting load schedules, as a fraction of peak lighting loads, in Figure 3-51 is correlated with the occupancy schedule. However, lighting has less variability than the occupancy schedule.

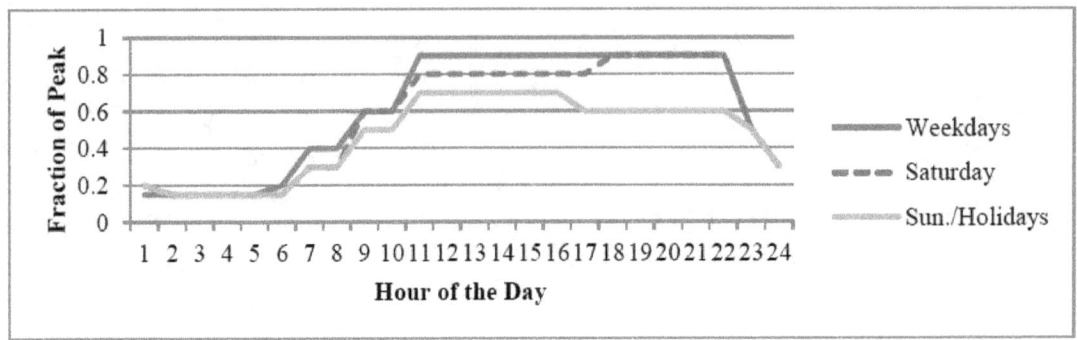

Figure 3-51 RSTRNT1 Lighting Schedule

The peak electrical equipment load is 502 W, or 1.08 W/m^2 (0.10 W/ft^2). The electrical load schedule in Figure 3-52 is highly correlated with the lighting schedules, but does not vary across days of the week because most electrical loads tend to be constant when a building is occupied no matter the amount of occupancy.

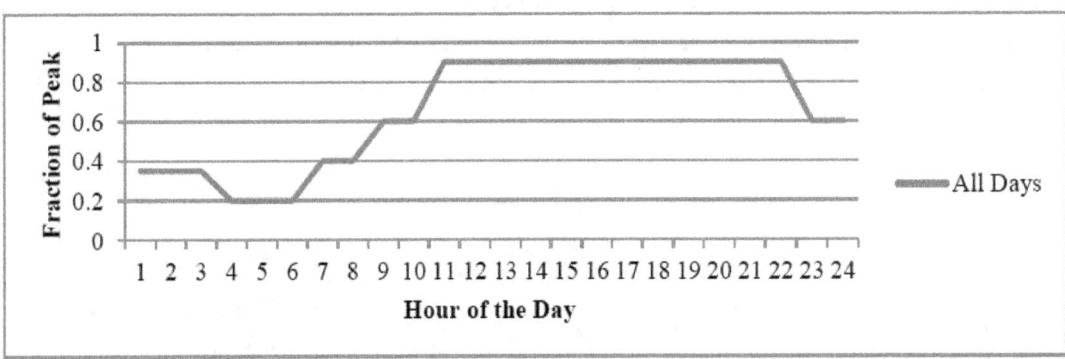

Figure 3-52 RSTRNT1 Electrical Load Schedule

4   Environmental Performance Measurement

BIRDS evaluates the environmental performance of whole buildings using a life cycle assessment (LCA) approach. The general LCA methodology involves four steps (ISO, 2006). The *goal and scope definition* step spells out the purpose of the study and its breadth and depth. The *inventory analysis* step identifies and quantifies the environmental inputs and outputs associated with a building over its entire life cycle. Environmental inputs include water, energy, land, and other resources; outputs include releases to air, land, and water. However, it is not these inputs and outputs, or *inventory flows,* which are of primary interest. We are more interested in their consequences, or impacts on the environment. Thus, the next LCA step, *impact assessment*, characterizes these inventory flows in relation to a set of environmental impacts. For example, the impact assessment step might relate carbon dioxide emissions, a *flow*, to global warming, an *impact*. Finally, the *interpretation* step combines the environmental impacts in accordance with the goals of the LCA study.

4.1   Goal and Scope Definition

The goal of BIRDS LCAs is to generate environmental performance scores for a range of U.S. building types, each designed to meet 5 alternative levels of operating energy efficiency. These results will be reported alongside economic performance scores to help designers, investors, and policymakers develop business cases for high-performance green buildings.

The scoping phase of any LCA involves defining the boundaries of the product system—or, in the case of BIRDS, the building—under study. In traditional bottom-up LCAs, boundary setting requires expert judgment by the analyst. The construction of a building involves a number of unit processes (e.g., asphalt production for input to the manufacture of facing for fiberglass batt insulation). Each unit process involves many inventory flows, some of which themselves involve other, subsidiary unit processes. Because including an ever-expanding number of unit processes in LCAs is not feasible, the product system's supply chain links are truncated at some point to include only those judged to make non-negligible contributions to the product system. The analyst typically uses mass, energy, and/or cost contributions as decision criteria. Use of different boundary setting criteria is one of the main reasons LCA results from different studies are often found incomparable.

One important advantage of the BIRDS hybrid approach is that it addresses the bottom-up issue of truncation in supply chain links, thereby improving consistency in system boundary selections. Through the hybridization process, truncated supply chain links are connected to the background U.S. economic system represented by the top-down data. These linkages follow the metabolic structure of the U.S. economy, thereby benefitting from a more complete system definition including potentially thousands of supply chain interactions.

Due to their long service lives, buildings are somewhat unique when it comes to the end-of-life stage of the life cycle. For most non-consumable product LCAs, end-of-life waste flows are included in the inventory analysis for full coverage of the life cycle. If there is an active recycling market that diverts some of the product from the waste stream, that portion of the product's end-of-life flows can be ignored. In BIRDS, however, building lifetimes range from 41 to 65 years, longer than the 40-year maximum length for the study period. Therefore, 100 % of each building is considered to be "recycled" at the end of the study period and there are no end-of-life waste flows allocated to the building at the end of the BIRDS study period. Rather, end-of-life waste flows should be allocated to a different "product," representing use of the building from the end of the study period to the end of the building service life. Similarly, the environmental burdens from building *construction* are allocated only to its first use (over the BIRDS study period); LCAs for all subsequent uses should be treated as free of these initial construction burdens. This effectively credits the use of existing buildings over new construction and ensures there will be no double counting once existing building LCAs are included in future versions of BIRDS as planned.

Defining the unit of comparison is an important task in the goal and scoping phase of LCA. The basis for all units of comparison is the *functional unit*, defined so that the systems compared are true substitutes for one another. **In the BIRDS model, the functional unit is construction and use of one building prototype over a user-defined study period.** The functional unit provides the critical reference point to which the LCA results are scaled.

Scoping also involves setting data requirements. With respect to geographic coverage, the BIRDS inventory data are generally U.S. average data. An exception is made for the electricity production inventory data applied to a building's use of electricity. These data are customized to each U.S. state. In terms of technology coverage, the top-down inventory data represent the mix of technologies in place as of 2002, the year of the most recent top-down data available from the U.S. Economic Census. For the bottom-up inventory data on building energy technologies, the most representative technology for which data are available is evaluated.

4.2 Life Cycle Inventory Analysis

BIRDS applies a hybridized life-cycle assessment approach. The approach is hybridized in the sense that a mixture of top-down and bottom-up data are collected and systematically integrated in the inventory analysis LCA step. Traditional process-based LCAs gather data by modeling all the in-scope industrial processes involved in a product's production (raw materials acquisition, materials processing, manufacture, transportation), use, and waste management. For each industrial process, the analyst collects very detailed, bottom-up data on all its inputs from the environment (e.g., materials, fuel, water, land) and outputs to the environment (e.g., products, water effluents, air emissions, waste). This process is summarized in Figure 4-1.

To address the complexities of a whole building, BIRDS takes a new, multi-layered approach to inventory analysis. Since a building's operating energy efficiency has an important influence on its sustainability performance, and energy efficiency is largely driven by the building's energy technologies, BIRDS pays special attention to the materials used in them. Specifically, BIRDS uses detailed life cycle inventory data for a range of energy technology packages that have been analyzed at the traditional, bottom-up LCA level. These energy technology packages are used to meet the 5 levels of energy efficiency simulated for each building type in 228 different U.S. locations. The bottom-up approach is also used to gather inventory data for a building's use of electricity and natural gas over the study period. These bottom-up BIRDS data were developed under contract to NIST by Four Elements Consulting, LLC, of Seattle, Washington, and are documented in section 4.5. For all other building constituents, industry average life cycle inventory data are gathered from the top down and then systematically combined with bottom up data into a comprehensive, hybrid life cycle inventory for a whole building.

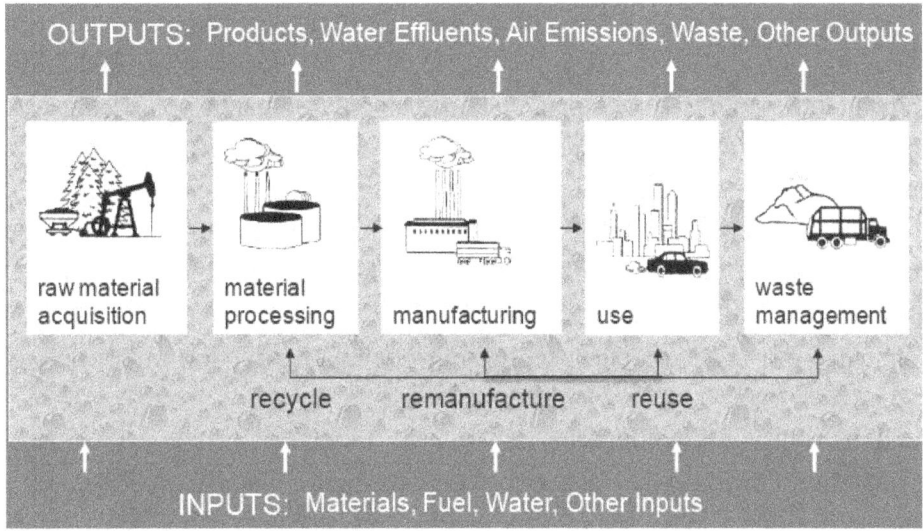

Figure 4-1. Compiling LCA Inventories of Environmental Inputs and Outputs

The inventory data items collected through the bottom up and top down approaches are identical—for example, kg carbon dioxide, kWh primary energy consumption—but some of the data sources are quite different. The systematic hybridization of the data sets bridges these differences to yield coherent and consistent BIRDS life cycle inventories for a wide variety of new commercial buildings. The LCAs for the buildings are then completed by applying conventional methods of life cycle impact assessment and interpretation to the hybrid inventory data.

Top-Down Inventory Analysis. An economy's accounting structure provides a cost-effective top-down approach to LCA inventory data collection. Many developed economies maintain economic input-output (I-O) accounts that trace the flow of goods and services throughout industries. Much the same way that a product's production can be traced upstream through its supply chain, an industry's production can be traced upstream through its supply chain. The U.S. Census Bureau conducts an Economic Census of U.S. industry every five years that does just that. Covering 97 % of business receipts, the census reaches nearly all U.S. business establishments. Based on the detailed data collected, the U.S. Bureau of Economic Analysis (BEA) creates what are called Input-Output Accounts, or "I-O tables," for the U.S. economy.

The U.S. I-O tables show how around 500 industries provide input to, and use output from, each other to produce Gross Domestic Product (GDP). These tables, for example, can show how $100 million of U.S. economic output in the residential building construction sector traces back through its direct monetary inputs—from the construction process itself—to its indirect inputs from contributing sectors such as the steel, concrete, lumber, and plastics industries. Economic output from the steel, concrete, lumber, and plastics industries, in turn, can be traced back through those supply chains such as mining, forestry, and fuel extraction. And so on.

While BEA provides these I-O tables in purely monetary terms, academics have successfully developed "environmentally-extended" I-O tables (Suh, 2005; Hendrickson, 2006; Suh, 2010). These top-down tables tap into a wide range of national environmental statistics to associate environmental inputs and outputs with economic activity in industry sectors, including use of raw materials, fuel, water, and land and releases of water effluents, air emissions, and waste. BIRDS uses environmentally-extended I-O tables for the U.S. construction industry developed under contract to NIST by Industrial Ecology Research Services of Goleta, California. These tables are based on the 2007 release of the 2002 BEA I-O data, the latest available, and quantify 6204 environmental inputs and outputs occurring throughout production supply chains.

To understand the contribution of building construction to the nation's environmental footprint (impact), it is useful to focus on the concept of "final demand." The BEA's monetary I-O tables use GDP—the total value of the consumption of goods and services in a year—to measure final demand. This value consists of spending and investment by consumers, businesses, and government, as well as net exports. Since final demand is satisfied through annual production— goods and services need to be produced before they can be bought—each industry's value-added, or "direct" contribution to GDP, reflects its share of final demand.

The environmentally-extended I-O tables translate economic activity into environmental terms, or monetary GDP into environmental GDP (eGDP). In LCA terms, the construction industry's contribution to eGDP is not limited to the direct impact from value-added construction processes and activities. Its contribution to eGDP also includes the indirect impacts stemming from contributions by upstream construction supply chains up to and including raw materials acquisition. The supply chain relationships built into the environmentally-extended I-O tables enable estimation of construction industry impacts on this cumulative, life-cycle basis. Figure 4-2 illustrates these supply chain relationships for some of the inputs to ready-mix concrete manufacturing, an indirect construction industry input.

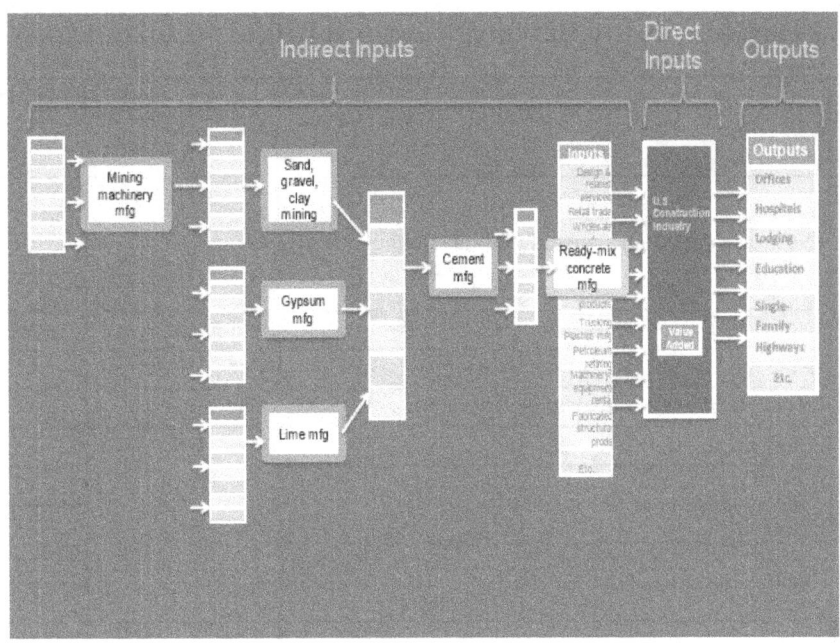

Figure 4-2. Illustration of Supply Chain Contributions to U.S. Construction Industry

The environmentally extended I-O tables classify U.S. construction into 42 distinct industry outputs. In this first version of BIRDS, top-down inventory data represent this level of detail for the construction, maintenance, and repair associated with the 9 industry outputs shown in Table 4-1. The first 7 outputs correspond to the occupancy types covered by the 11 building prototypes in BIRDS. The last 2 correspond to maintenance and repair (M&R) activities in those buildings. For routine M&R, nonresidential M&R output applies to all but the lodging occupancy prototypes. For these, residential M&R output applies. For all the construction industry outputs, the baseline top-down inventory data are expressed in terms of life-cycle environmental flows *per dollar* of construction.

Table 4-1. Construction Industry Outputs Mapped to BIRDS Building Types

Construction Type	Industry Output	Occupancy	BIRDS Building Type
New Construction	New office buildings, including financial buildings	Office	OFFIC03
			OFFIC08
			OFFIC16
	New multi-merchandise shopping	Mercantile	RETAIL1
	New food and beverage establishments	Food Service	RSTRNT1
	New educational and vocational structures	Education	HIGHS02
	New lodging	Lodging	HOTEL15
	New multifamily residential structures		APART04
			APART06
	New dormitories		DORMI04
			DORMI06
M&R Construction	Residential maintenance and repair construction	Lodging	HOTEL15
			APART04
			APART06
			DORMI04
			DORMI06
	Nonresidential maintenance and repair construction	All Others	All Others

One advantage of the BIRDS approach is the economic dimension built into the top-down inventory data. These data are directly associated with U.S. economic data, permitting seamless integration of the economic dimension in the BIRDS sustainability measurement system. The top-down inventory values on a per-dollar basis are multiplied by the corresponding BIRDS construction, maintenance, and repair costs to translate them into the LCA functional unit representing the whole building over a user-defined study period.

For more information on the mathematics, accounting structure, and step-by-step process under which the BIRDS hybrid environmental database is built, see Suh and Lippiatt 2012.

4.3 Life Cycle Impact Assessment

Environmental impacts from building construction and use derive from the 6204 inputs and outputs occurring throughout production supply chains, as quantified in the hybrid BIRDS life cycle inventory. The impact assessment step of LCA quantifies the potential contribution of these inventory items to a range of environmental impacts. The approach preferred by most LCA practitioners and scientists today involves a two-step process:

- *Classification* of inventory flows that contribute to specific environmental impacts. For example, greenhouse gases such as carbon dioxide, methane, and nitrous oxide are classified as contributing to global warming.
- *Characterization* of the potential contribution of each classified inventory flow to the corresponding environmental impact. This results in a set of indices, one for each impact, which is obtained by weighting each classified inventory flow by its relative contribution to the impact. For instance, the Global Warming Potential index is derived by expressing each greenhouse gas in terms of its equivalent amount of carbon dioxide heat trapping potential.

There are two general applications of this life cycle impact assessment (LCIA) approach: midpoint-level and endpoint-level analyses. An endpoint-level analysis attempts to measure the ultimate damage that each environmental input and output in the inventory will have along the cause-effect chain. Methods of this type include just a few impact categories, such as damage to human health, ecosystems, and resource availability, that are easier to interpret in the final step of life cycle assessment. This approach is criticized for the numerous assumptions, value judgments, and gaps in coverage of the underlying damage models. A midpoint-level analysis, on the other hand, selects points along the cause-effect chain at which more certain and comprehensive assessments may be carried out. While this approach generates many impact categories, making life-cycle interpretation more difficult, it is more scientifically defensible. Even so, a midpoint-level analysis does not offer the same degree of relevance for all environmental impacts. For global and regional effects (e.g., global warming and acidification) the method provides an accurate description of the potential impact. For impacts dependent upon local conditions (e.g., smog), it may result in an oversimplification of the actual impacts because the indices are not tailored to localities. Note that some impact assessments apply a mix of midpoint and endpoint approaches.

4.3.1 BIRDS Impact Assessment

BIRDS uses a midpoint-level analysis to translate its 6204 environmental inputs and outputs into a manageable set of science-based measurements across 12 environmental impacts. BIRDS primarily uses the U.S. Environmental Protection Agency's TRACI (Tool for the Reduction and Assessment of Chemical and other environmental Impacts) version 2.0 set of state-of-the-art, peer-reviewed U.S. life cycle impact assessment methods (Bare 2011). Since TRACI 2.0 does not include land and water use, these two important resource depletion impacts are assessed using other characterization methods (Guinee 2002, Goedkoop 2009). Together these methods are used to develop BIRDS performance metrics indicating the degree to which construction and use of a building contributes to each environmental impact. Following are brief descriptions of the 12 BIRDS impact categories.

Impact Categories

Global Warming. The Earth absorbs radiation from the Sun, mainly at the surface. This energy is then redistributed by the atmosphere and ocean and re-radiated to space at longer wavelengths. Some of the thermal radiation is absorbed by "greenhouse" gases in the atmosphere, principally water vapor, but also carbon dioxide, methane, the chlorofluorocarbons, and ozone. The absorbed energy is re-radiated in all directions, downwards as well as upwards, such that the radiation that is eventually lost to space is from higher, colder levels in the atmosphere. The result is that the surface loses less heat to space than it would in the absence of the greenhouse gases and consequently stays warmer than it would be otherwise. This phenomenon, which acts rather like a 'blanket' around the Earth, is known as the greenhouse effect.

The greenhouse effect is a natural phenomenon. The environmental issue is the change in the greenhouse effect due to emissions (an increase in the effect) and absorptions (a decrease) attributable to humans. A general increase in temperature can alter atmospheric and oceanic temperatures, which can potentially lead to alteration of circulation and weather patterns. A rise in sea level is also predicted from an increase in temperature due to thermal expansion of the oceans and melting of polar ice sheets.

Primary Energy Consumption. Primary energy consumption leads to fossil fuel depletion when fossil fuel resources are consumed at rates faster than nature renews them. Some experts believe fossil fuel depletion is fully accounted for in market prices. That is, market price mechanisms are believed to take care of the scarcity issue, price being a measure of the level of depletion of a resource and the value society places on that depletion. However, price is influenced by many factors other than resource supply, such as resource demand and non-perfect markets (e.g., monopolies and subsidies). The primary energy consumption metric is used to account for the resource depletion aspect of fossil fuel extraction.

Human Health—Criteria Air Pollutants can arise from many activities including combustion, vehicle operation, power generation, materials handling, and crushing and grinding operations. They include coarse particles known to aggravate respiratory conditions such as asthma, and fine particles that can lead to more serious respiratory symptoms and disease.

Human Health—Cancer Effects can arise from exposure to industrial and natural substances, and can lead to illness, disability, and death. Its assessment is based on the global consensus model known as USEtox, which describes the fate, exposure and effects of thousands of chemicals.

Water Consumption. Water resource depletion has not been routinely assessed in LCAs to date, but researchers are beginning to address this issue to account for areas where water is scarce, such as the Western United States. While some studies use water *withdrawals* to evaluate this impact, a more refined analysis considers that a portion of water withdrawn may be returned through evapotranspiration (the sum of evaporation from surface water, soil, and plant leaves). BIRDS uses the latter approach to measure water *consumption*, or water withdrawn net of evapotranspiration. BIRDS evaluates water consumption from cradle to grave, including water consumption during building use.

Ecological Toxicity measures the potential of pollutants from industrial sources to harm land- and water-based ecosystems. Its assessment is based on the global consensus model known as USEtox, which describes the fate, exposure and effects of thousands of chemicals.

Eutrophication Potential. Eutrophication is the addition of mineral nutrients to the soil or water. In both media, the addition of large quantities of mineral nutrients, such as nitrogen and phosphorous, results in generally undesirable shifts in the number of species in ecosystems and a reduction in ecological diversity. In water, it tends to increase algae growth, which can lead to lack of oxygen and therefore death of species like fish.

Land Use. This impact category measures the use of land resources by humans which can lead to undesirable changes in habitats. Note that the BIRDS land use approach does not consider the original condition of the land, the extent to which human activity changes the land, or the length of time required to restore the land to its original condition. As impact assessment science continues to evolve, it is hoped that these potentially important factors will become part of BIRDS land use assessment.

Human Health—Noncancer Effects can arise from exposure to industrial and natural substances, and range from transient irritation to permanent disability and even death. Its assessment is based on the global consensus model known as USEtox, which describes the fate, exposure and effects of thousands of chemicals.

Smog Formation. Smog forms under certain climatic conditions when air emissions from industry and transportation are trapped at ground level where they react with sunlight. Smog leads to harmful impacts on human health and vegetation.

Acidification Potential. Acidifying compounds may in a gaseous state either dissolve in water or fix on solid particles. They reach ecosystems through dissolution in rain or wet deposition. Acidification affects trees, soil, buildings, animals, and humans. The two compounds principally involved in acidification are sulfur and nitrogen compounds. Their principal human source is fossil fuel and biomass combustion. Other compounds released by human sources, such as hydrogen chloride and ammonia, also contribute to acidification.

Ozone Depletion, or a thinning of the stratospheric ozone layer, allows more harmful short wave radiation to reach the Earth's surface, potentially causing undesirable changes in ecosystems, agricultural productivity, skin cancer rates, and eye cataracts, among other issues.

Computational Algorithms
There are six building components represented in the BIRDS life-cycle impact assessment (LCIA) calculations for whole buildings:

- Baseline building: new construction (Base_New)
- Baseline building: maintenance and repair over study period (Base_M&R)
- Energy technology package: new construction (ETP_New)
- Energy technology package: annual maintenance and repair (ETP_M&R)
- Annual operating energy use: electricity (ELEC)
- Annual operating energy use: natural gas (NG)

The hybridized life cycle inventory data for each component are expressed in different units For example, the baseline building inventories are given on a per-dollar basis, the energy technology package inventories on a per-physical unit basis (usually area), and the operating energy use inventories on a per-BTU basis. Thus, each requires its own LCIA computational algorithm as shown in Table 4-2. These calculations ensure that after adjusting for study period length, all LCIA results are expressed in the consistent functional unit defined for BIRDS: construction and use of one building prototype over a user-defined study period.

Table 4-2 BIRDS Life Cycle Impact Assessment Calculations by Building Component

Building Component	LCIA Equation	Notation
Base_New	$LCIA_{i,j,c=1}$ $=$ $(LCIA_{i,j,c=1}/\$)*\$_{i,c=1}$	LCIA=classified and characterized life cycle inventories
Base_M&R	$LCIA_{i,j,c=2,yr}$ $=$ $(LCIA_{i,j,c=2}/\$)*\$_{i,c=2,yr}$	c=construction type code, 1=new, 2=M&R E=electricity ET=energy technology product
ETP_New	$LCIA_{i,j,c=1,x,K,T}$ $=$ $\Sigma (LCIA_{j,c=1,ET(i,x,K,T)}/FU)*FU_{i,T}$ from T=1 to 6	FU=functional unit* i = building type, i=1 to 11
ETP_M&R	$LCIA_{i,j,c=2,x,K,T}/yr$ $=$ $\Sigma (LCIA_{j,c=2,ET(i,x,K,T)}/FU/yr)*FU_{i,T}$ from T=1 to 6	j=environmental impact, j=1 to 12 K=energy standard, K=1 to 5 n=study period length in years, n=1 to 40
ELEC	$LCIA_{i,j,x,K}/yr$ $=$ $(LCIA_{j,s}/BTU_E)*(BTU_{E,i,x,K}/yr)$	NG=natural gas s=U.S. state, s=1 to 50
NG	$LCIA_{i,j,x,K}/yr$ $=$ $(LCIA_j/BTU_{NG})*(BTU_{NG,i,x,K}/yr)$	T = energy technology group, T=1 to 6* x= building location, x=1 to 228

*energy technology groups and their functional units are wall insulation (ft^2), roof insulation (ft^2), windows (ft^2), HVAC (no. of units), overhangs (ft^2), and daylighting (ft^2).

4.3.2 BIRDS Normalization

Once impacts have been classified and characterized, the resulting LCIA metrics are expressed in incommensurate units. Global warming is expressed in carbon dioxide equivalents, acidification in hydrogen ion equivalents, eutrophication in nitrogen equivalents, and so on. In order to assist in the next LCA step, interpretation, these metrics are often placed on the same scale through normalization.

The U.S. EPA has developed "normalization references" corresponding to its TRACI set of impact assessment methods (Bare et al. 2006). These U.S. data are updated and expanded for use in BIRDS. Shown in Table 4-3, these values quantify the U.S. economy's annual contributions to each impact category. As such, they represent a "U.S. impact yardstick" against which to evaluate the *significance* of building-specific impacts. Normalization is accomplished by dividing BEES building-specific impact assessment results by the fixed U.S.-scale normalization references, expressed in the same units, yielding an impact category score for a building that has been placed in the context of annual U.S. contributions to that impact. By placing each building-specific impact result in the context of its associated U.S. impact result, the measures are all reduced to the same scale, allowing comparison across impacts.

Table 4-3. BIRDS Normalization References

Impact Category	Normalization reference (U.S. total/yr)	Units
Global Warming	7.16E+12	kg carbon dioxide equivalents
Primary Energy Consumption	3.52E+13 (1.20E+14)	kWh (kBTU)
HH Criteria Air	2.24E+10	kg particulate matter 10 equivalents (PM10 = particulate matter <10 microns)
HH Cancer	1.05E+04	comparative human toxicity units
Water Consumption	1.69E+14	L
Ecological Toxicity	3.82E+13	comparative ecotoxicity units
Eutrophication	1.01E+10	kg nitrogen equivalents
Land Use	7.32E+08 (1.81E+09)	hectare (acre)
HH Noncancer	5.03E+05	comparative human toxicity units
Smog Formation	4.64E+11	kg ozone equivalents
Acidification	1.66E+12	moles hydrogen ion equivalents
Ozone Depletion	5.10E+07	kg CFC-11 equivalents (CFC-11 = trichlorofluoromethane)

Normalized BIRDS impact scores have powerful implications. By evaluating a building's impacts with reference to their importance in a larger context, an impact to which one building design contributes little will not appear important when, by comparison, competing designs contribute even less to that impact.

4.4 Life Cycle Interpretation

At the BIRDS LCA interpretation step, a building's normalized impact scores are evaluated. The midpoint-level impact assessment yields scores for twelve impact categories, making interpretation at this level difficult. To enable comparisons across buildings, the scores across impact categories may be synthesized. Note that in BIRDS, synthesis of impact scores is optional.

Impact scores may be synthesized by weighting each impact category by its relative importance to overall environmental performance, then computing the weighted average impact score. In the BIRDS software, the set of importance weights is selected by the user. Several alternative weight sets are provided as guidance, and may be either used directly or as a starting point for developing user-defined weights. The alternative weights sets are based on an EPA Science Advisory Board study, a BEES Stakeholder Panel's structured judgments, a set of equal weights, and a set exclusively focusing on the global warming impact, representing a spectrum of ways in which people value diverse aspects of the environment.

4.4.1   EPA Science Advisory Board study

In 1990 and again in 2000, EPA's Science Advisory Board (SAB) developed lists of the relative importance of various environmental impacts to help EPA best allocate its resources. (U.S. EPA 1990, U.S. EPA 2000). The following criteria were used to develop the lists:

- The spatial scale of the impact
- The severity of the hazard
- The degree of exposure
- The penalty for being wrong

Ten of the twelve BIRDS impact categories were covered by the SAB lists of relative importance:

- Highest-Risk Problems: global warming, land use
- High-Risk Problems: ecological toxicity, human health (cancer and noncancer effects)
- Medium-Risk Problems: ozone depletion, smog, acidification, eutrophication, human health—criteria air pollutants

The SAB did not explicitly consider primary energy consumption or water consumption. For this exercise, these impacts are assumed to be relatively medium-risk and low-risk problems, respectively, based on other relative importance lists. (Levin 1996).

Verbal importance rankings, such as "highest risk," may be translated into numerical importance weights by following ASTM International standard guidance for applying a Multiattribute Decision Analysis method known as the Analytic Hierarchy Process. (ASTM, 2011). The AHP methodology suggests the following numerical comparison scale:

1	Two impacts contribute equally to the objective (in this case environmental performance)
3	Experience and judgment slightly favor one impact over another
5	Experience and judgment strongly favor one impact over another
7	One impact is favored very strongly over another, its dominance demonstrated in practice
9	The evidence favoring one impact over another is of the highest possible order of affirmation

2, 4, 6, and 8 can be selected when compromise between values of 1, 3, 5, 7, and 9, is needed.

Through an AHP process known as pairwise comparison, numerical comparison values are assigned to each possible pair of environmental impacts. Relative importance weights can then be derived by computing the normalized eigenvector of the largest eigenvalue of the matrix of pairwise comparison values. Table 4-4 and Table 4-5 list the pairwise comparison values assigned to the verbal importance rankings, and the resulting SAB importance weights computed for the BIRDS impacts, respectively. Note that the pairwise comparison values were assigned through an iterative process based on NIST's background and experience in applying the AHP technique. Furthermore, while the SAB evaluated cancer and noncancer effects as a group, the resulting 13 % weight was apportioned between the two based on the relative judgments of the BEES Stakeholder Panel discussed in the next section.

Table 4-4. Pairwise Comparison Values for Deriving Impact Category Importance Weights

Verbal Importance Comparison	Pairwise Comparison Value
Highest vs. Low	6
Highest vs. Medium	3
Highest vs. High	1.5
High vs. Low	4
High vs. Medium	2
Medium vs. Low	2

Table 4-5. Relative Importance Weights based on Science Advisory Board Study

Impact Category	Relative Importance Weight (%)
Global Warming	18
Primary Energy Consumption	7
HH Criteria Air	7
HH Cancer	8
Water Consumption	3
Ecological Toxicity	12
Eutrophication	5
Land Use	18
HH Noncancer	5
Smog Formation	7
Acidification	5
Ozone Depletion	5

4.4.2   BEES Stakeholder Panel judgments

While the derived EPA SAB-based weight set is helpful and offers expert guidance, several interpretations and assumptions were required in order to translate SAB findings into numerical weights for interpreting LCA-based analyses. A more direct approach to weight development would consider a closer match to the context of the application; that is, environmentally preferable purchasing in the United States based on life-cycle impact assessment results, as reported by BIRDS.

In order to develop such a weight set, NIST assembled a volunteer stakeholder panel that met at its facilities in Gaithersburg, Maryland, for a full day in May 2006. To convene the panel, invitations were sent to individuals representing one of three "voting interests:" producers (e.g., building product manufacturers), users (e.g., green building designers), and LCA experts. Nineteen individuals participated in the panel: seven producers, seven users, and five LCA experts. These "voting interests" were adapted from the groupings ASTM International employs for developing voluntary standards, in order to promote balance and support a consensus process.

The BEES Stakeholder Panel was led by Dr. Ernest Forman, founder of the AHP firm Expert Choice Inc. Dr. Forman facilitated panelists in weighting the BEES impact categories using the AHP pairwise comparison process. The panel weighted all impacts in the Short Term (0 years to 10 years), Medium Term (10 years to 100 years), and Long Term (>100 years). One year's worth of U.S. flows for each pair of impacts was compared, with respect to their contributions to environmental performance. For example, for an impact comparison over the Long Term, the panel was evaluating the effect that this year's U.S. emissions would have more than 100 years hence.

Once the panel pairwise-compared impacts for the three time horizons, its judgments were synthesized across these time horizons. Note that when synthesizing judgments across voting interests and time horizons, all panelists were assigned equal importance, while the short, medium, and long-term time horizons were assigned by the panel to carry 24 %, 31 %, and 45 % of the weight, respectively.

The environmental impact importance weights developed through application of the AHP technique at the facilitated BEES Stakeholder Panel event are shown in Table 4-6. These weights reflect a synthesis of panelists' perspectives across all combinations of stakeholder voting interest and time horizon. The weight set draws on each panelist's personal and professional understanding of, and value attributed to, each impact category. While the synthesized weight set may not equally satisfy each panelist's view of impact importance, it does reflect contemporary values in applying LCA to real world decisions. This synthesized BEES Stakeholder Panel weight set is offered as an option in BIRDS online.

The panel's application of the AHP process to derive environmental impact importance weights is documented in an appendix to ASTM Standard E1765-11 and in Gloria et al. 2007.

Table 4-6. Relative Importance Weights based on BEES Stakeholder Panel Judgments

Impact Category	Relative Importance Weight (%)
Global Warming	29.9
Primary Energy Consumption	10.3
HH Criteria Air	9.3
HH Cancer	8.2
Water Consumption	8.2
Ecological Toxicity	7.2
Eutrophication	6.2
Land Use	6.2
HH Noncancer	5.2
Smog Formation	4.1
Acidification	3.1
Ozone Depletion	2.1

Note: Since BIRDS does not currently include an Indoor Air Quality impact category, its 3 % BEES Stakeholder Panel weight has been redistributed among the remaining 12 impacts.

The three figures below display in graphical form the BEES Stakeholder Panel weights used in BIRDS. Figure 4-3 displays the synthesized weight set, Figure 4-4 the weights specific to panelist voting interest, and Figure 4-5 the weights specific to time horizon. The BIRDS user is free to interpret results using either of the weight sets displayed in Figure 4-4 and Figure 4-5 by entering them as a user-defined weight set.

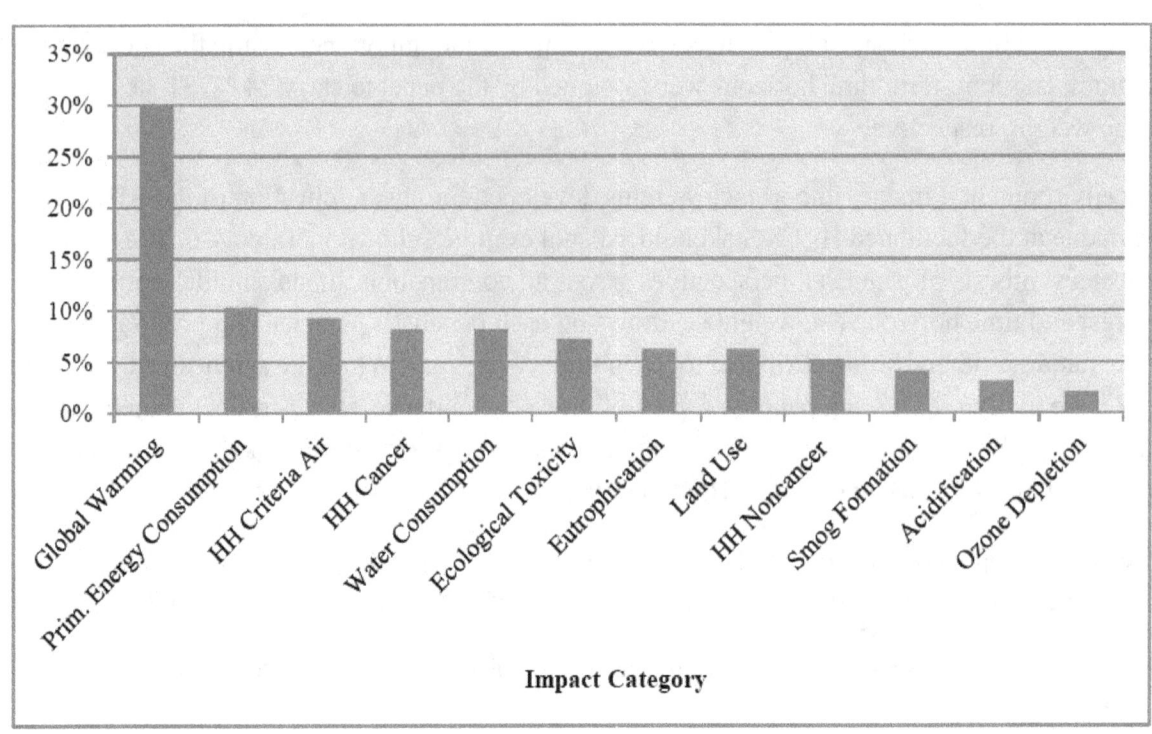

Figure 4-3. BEES Stakeholder Panel Importance Weights Synthesized across Voting Interest and Time Horizon

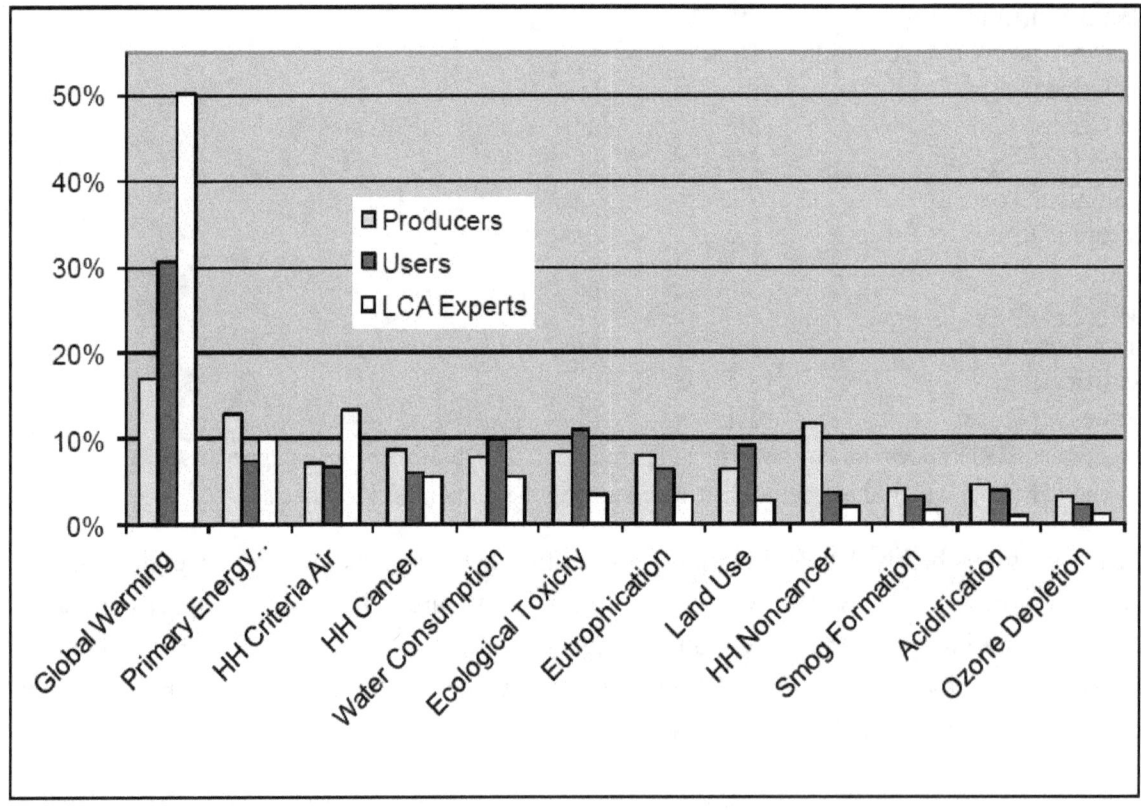

Figure 4-4. BEES Stakeholder Panel Importance Weights by Stakeholder Voting Interest

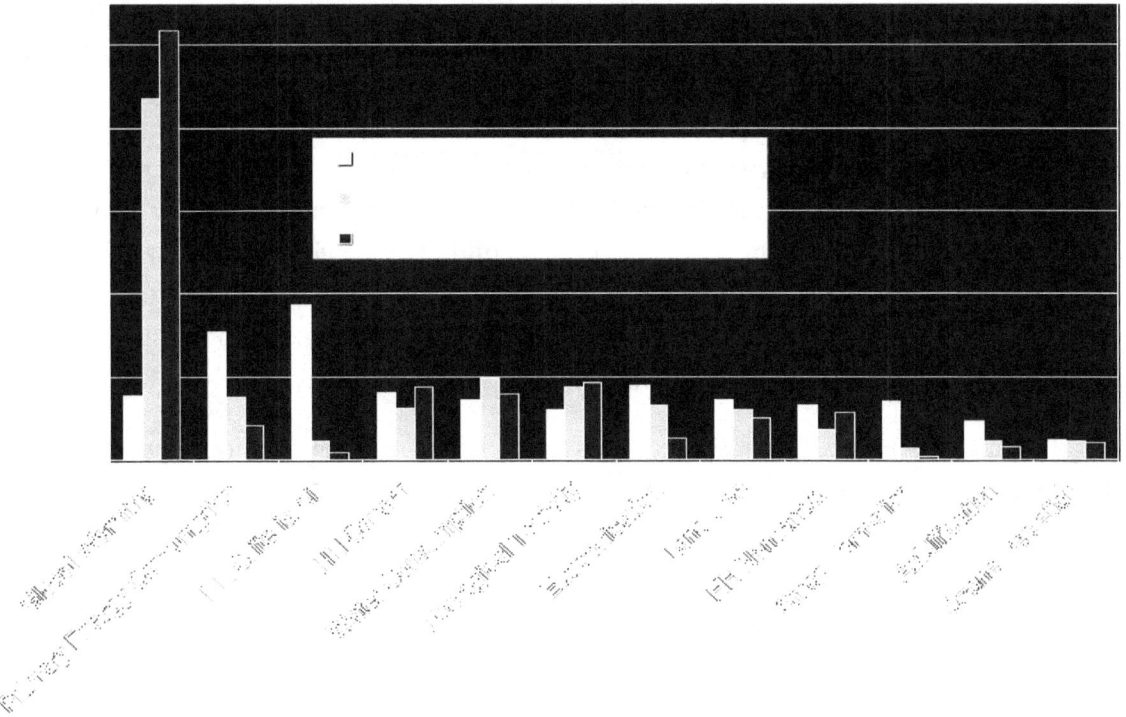

Figure 4-5. BEES Stakeholder Panel Importance Weights by Time Horizon

4.5 BIRDS Energy Technologies

Since buildings have very long lives, operating energy efficiency has an important influence on their sustainability performance. Energy efficiency is largely driven by a building's energy technologies, but top-down inventory data are not readily available at this level of resolution. Therefore, BIRDS includes detailed life cycle inventory data for a range of energy technology packages that have been compiled at the traditional, bottom-up LCA level. These energy technology packages are used to meet the 5 levels of energy efficiency simulated for each of 11 building types in 228 different U.S. locations. The bottom-up BIRDS data were developed under contract to NIST by Four Elements Consulting, LLC, of Seattle, Washington. Energy technologies include wall and roof insulation, windows, HVAC systems, overhangs, and daylighting.[6]

---

[6] Besides daylighting, the "Low Energy Case" building design decreases lighting density by first increasing the efficiency of the lighting system and then decreasing the number of fixtures in the lighting system. These actions are thought to qualitatively offset one another and so are not included in the life-cycle inventory analysis in this first version of BIRDS.

4.5.1 General Information Regarding the Energy Technology LCIs

Standards Used

The LCAs in BIRDS have been built based on the principles and framework in the International Organization for Standardization (ISO) 14040 (2006) and the guidelines specified in ISO 14044 (2006).

Primary and Secondary Data Sources

Both primary data (collected directly from a manufacturing facility) and secondary data (publicly-available literature sources) can be used to build LCAs, and it is common to see a combination of both data types, based on data availability. Sources of data on the energy technologies in BIRDS vary from one category to the next, and within categories themselves for the different products. Data were based on:

- Primary data from a group of companies and/or an industry association, compiled into an industry average product;
- Primary data on a product provided by one company;
- Secondary data that represent an average or typical product; and
- Secondary data that represent one product in a category.

For optimal data quality of an LCA, the preference is to have the most representative data – temporally, technologically, and geographically – on a product or system, so that the model produced most closely represents the product. But this is often not possible to achieve due to data availability constraints. It is also not always possible to have a data set that represents an entire category of products. For example, high quality, current, company-specific data might be collected and used to build the LCA for a given product. Likewise, data for another product might be compiled from literature sources due to lack of other available data. In both cases, the LCI profiles may be used to represent the full product category, even though they may not be representative of all products within the category, based on market share, technology, geographical source, etc. As a result, the reader should be aware of this limitation.

Data Sources Used for the Background Data

Secondary data have been applied to production of material inputs, production and combustion of fuels used for process energy, and transportation processes. The U.S. LCI Database (NREL, 2005-present, hereinafter referred to as "U.S. LCI Database") and the ecoinvent Data v.2.2 database (ecoinvent, 2007) are the main sources of background data throughout the various life cycle stages. Other sources of data are used where data were not available from U.S. LCI Database or ecoinvent, and/or where they were deemed to be of better quality than these latter sources.

The following subsections describe modeling, assumptions, and data sources of the product life cycle data. Data for the production of material inputs for each product are described in the subchapters since these may vary for different industries. The following data aspects are consistent for all products except where noted differently in the text:

- All energy production, including production of fuels and conversion into energy and electricity production come from the U.S. LCI Database.
- All transportation data come from the U.S. LCI Database.
- Where ecoinvent or other non-North American data sets were used, they were customized into North American processes by switching out foreign energy, electricity, transportation, and other processes for comparable North American based data sets from the U.S. LCI databases. Exceptions to this are noted.

4.5.2 Wall and Roof Insulation

The insulation categories considered for the commercial building walls and roof are presented in the tables below along with the R-values needed to meet the necessary thickness of insulation products to meet the requirements of the building design (e.g., edition of *ASHRAE 90.1* or LEC). Characteristics of each insulation type, including density and R-values ($\frac{m^2 * K}{W}$ per cm or $\frac{ft^2 * K * h}{Btu}$ per inch) are presented in the specific products' subsections.

Table 4-7 Specified Insulation Types and R-Values – Wall Application

Wall Insulation type	R-value specified by NIST per-cm (per-inch)
Kraft faced fiberglass blanket	4.33 (11.00)
Kraft faced fiberglass blanket	5.12 (13.00)
Kraft faced fiberglass blanket	5.91 (15.00)
Kraft faced fiberglass + polyiso foam board	7.44 (18.90)
Kraft faced fiberglass + polyiso foam board	7.68 (19.50)
Kraft faced fiberglass + polyiso foam board	8.03 (20.40)
Kraft faced fiberglass + XPS foam board	9.84 (25.00)
Kraft faced fiberglass + XPS foam board	11.81 (30.00)
Kraft faced fiberglass + XPS foam board	13.78 (35.00)

Table 4-8 Specified Insulation Types and R-Values – Roof Application

Roof Insulation type	R-value specified by NIST per-cm (per-inch)
EPS with perlite	4.12 (10.47)
XPS foam board	5.91 (15.00)
XPS foam board	7.87 (20.00)
XPS foam board	9.84 (25.00)
XPS foam board	11.81 (30.00)
XPS foam board	13.78 (35.00)
XPS foam board	15.75 (40.00)
XPS foam board	17.72 (45.00)
XPS foam board	19.69 (50.00)
XPS foam board	21.65 (55.00)
XPS foam board	23.62 (60.00)
Polyiso foam board	2.01 (5.10)
Polyiso foam board	2.81 (7.14)
Polyiso foam board	4.28 (10.87)
Polyiso foam board	5.63 (14.29)
Polyiso foam board	6.56 (16.67)
Polyiso foam board	8.56 (21.74)
Polyiso foam board	9.84 (25.00)

BIRDS performance data for the insulation category was provided on the basis of 0.09 m^2 (1 ft^2) of the specified R-value of insulation, which was then multiplied by the needed amount of square area for each building. The flow diagram below presents the general system boundaries for the insulation category, as it is modeled for BIRDS.

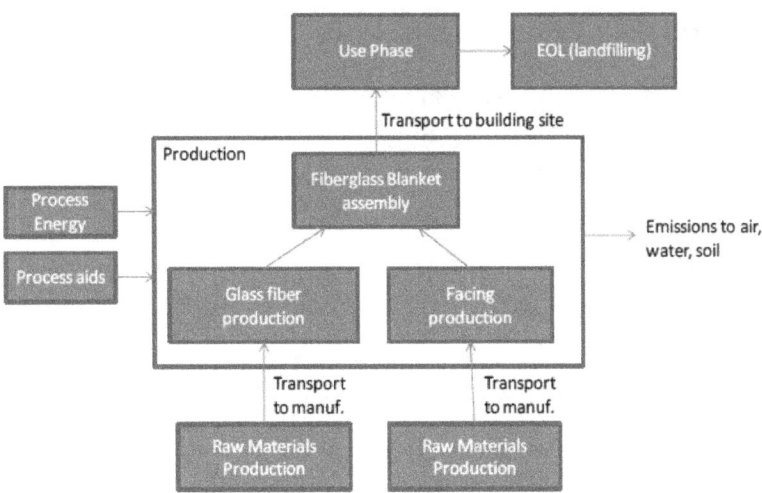

Figure 4-6 Insulation System Boundaries – Fiberglass Blanket Example

The modeling and assumptions for each type of insulation are presented below.

Fiberglass Blanket

Thermal efficiencies of R-11, R-13, and R-15 are used for wall insulation. R-15 is also used in the wall with polyisocyanurate and extruded polystyrene (XPS) foam boards. The table below specifies fiberglass insulation by type and R-value:

Table 4-9 Fiberglass Blanket Mass by Application

Application	Thickness	Density	Mass per Functional Unit
	cm (in)	kg/m^3 (lb/ft^3)	kg/m^2 (oz/ft^2)
Wall--R-11	8.9 (3.5)	12.1 (0.76)	1.07 (3.52)
Wall--R-13	8.9 (3.5)	12.1 (0.76)	1.07 (3.52)
Wall--R-15	8.9 (3.5)	22.6 (1.41)	2.01 (6.58)

Upstream Materials Production through Manufacturing

Fiberglass insulation is made with a blend of sand, limestone, soda ash, and recycled glass cullet. Recycled window, automotive, or bottle glass is increasingly used in the manufacture of glass fiber, and it accounts for 30 % to 50 % of the raw material input. The recycled content is limited by the amount of usable recycled material available in the market – not all glass cullet is of sufficient quality to be used in the glass fiber manufacturing process. The use of recycled material has helped to steadily reduce the energy required to produce insulation products. The raw materials used to produce fiberglass insulation are broken down by the glass and facing contents, shown in Table 4-10.

Table 4-10 Fiberglass Insulation Constituents

Glass Constituent	Mass Fraction (%)
Soda Ash	9
Borax	12
Glass Cullet	34
Limestone	9
Phenolic resin (binder coating)	5
Sand	31

Facing	Mass Fraction (%)
Kraft paper	25
Asphalt	75

The production data for the soda ash, limestone, and phenol formaldehyde resin come from the U.S. LCI Database. The borax, glass cullet, and silica sand come from ecoinvent. For the facing, Kraft paper comes from ecoinvent and the asphalt comes from U.S. LCI Database.

The raw materials are transported to the manufacturing plant via diesel truck. Materials are sourced domestically, and transportation distances range on average from 161 km (100 mi) to 805 km (500 mi).

The energy requirements for melting the glass constituents into fibers and drying of the completed blanket involve use of natural gas and electricity, shown in the following Table.

Table 4-11 Energy Requirements for Fiberglass Insulation Manufacturing

Energy Carrier	MJ/kg (Btu/lb)
Natural Gas	1.99 (857)
Electricity	1.37 (591)
Total	**3.36 (1 448)**

Besides combustion emissions from fuel usage at manufacturing, additional emissions are listed in the Table below.

Table 4-12 Non Fuel Combustion-Related Emissions for Fiberglass Insulation Manufacturing

Emission	Bonded Blankets g/kg (lb/ton)
Particulates	2.380 (4.759)
VOC	0.759 (1.518)

All waste produced during the cutting and blending process is either recycled into other insulation materials or added back into the glass mix. Thus, no solid waste is assumed to be generated during the production process.

Transportation to the Building Site through End of Life

Transportation of the insulation to the building site is modeled by an assumed average of 805 km (500 mi) by heavy-duty diesel-fueled truck.

Installing fiberglass blanket insulation is primarily a manual process; no energy or emissions are included in the model for this. During installation, any waste material is added into the building shell where the insulation is installed - there is effectively no installation waste.

Fiberglass insulation has a functional lifetime of over 50 years so no replacement is needed during the 40 year study period. How this product affects operating energy during the prototype buildings' use phase is addressed in other sections of this report.

While fiberglass insulation is recyclable, it is assumed that it is disposed of in a landfill at end of life. End-of-life modeling includes transportation by heavy-duty diesel-fuel powered truck approximately 80 km (50 mi) to a construction & demolition (C&D) landfill. Insulation in a landfill is modeled based on ecoinvent end of life waste management process data.

EPS Foam Insulation

Expanded polystyrene foam is rigid board stock used to provide wall, ceiling, roof, foundation perimeter and sub-slab insulation. EPS is also used in construction application systems such as exterior insulation finishes (EIFS), insulating concrete forms (ICF), and structural insulated panels (SIP). These applications combine EPS foam with a stucco finish, a poured concrete core or OSB or metal skins over an EPS core.

ASTM C-578 recognizes EPS foam as Type I, II, VIII, IX, XI, XIV, and XV. Densities of the various Types range from 11 kg/m^3 to 48 kg/m^3 (0.7 lb/ft^3 to 3.0 lb/ft^3) and R-values range from 1.22 per cm to 1.69 per cm (3.1 per inch to 4.3 per inch). Types I, VIII, II, and IX are the predominant types used for construction applications and deliver R values of at least (1.42, 1.50, 1.57, and 1.65) per cm ((3.6, 3.8, 4.0 and 4.2) per inch) at densities of (14.4, 18.4, 21.6, and 28.8) kg/m^3 ((0.90, 1.15, 1.35 and 1.80) lb/ft^3), respectively.

Upstream Materials Production through Manufacturing

Cradle-to-gate data on production through manufacturing is based on a 2009 study by Franklin Associates, A Division of ERG (Eastern Research Group) for the EPS Molders Association (Franklin, 2009). The scope of this study included collecting 2008 production data from 10 manufacturing facilities in the U.S. and Canada and compiling it into an industry average EPS foam board. The resulting cradle-to-gate unit process data has been published in U.S. LCI

Database. The following table presents the raw material inputs associated with the production of EPS insulation board (NREL 2005-present).

Table 4-13 EPS Foam Board Raw Materials

Raw Material Input	kg/kg foam (lb/lb foam)
Ethylbenzene/styrene	1.01 (1.01)
Processed natural gas	0.043 (0.043)
Petroleum	0.029 (0.029)

Production data for all inputs come from the U.S. LCI Database. Other inputs may be present in smaller quantities, and these may include flame retardant and/or other additives. No data were provided on these. The raw materials are transported to the manufacturing plant via diesel truck or pipeline. Materials are sourced domestically, and transportation distances range on average from 161 km (100 mi) to 805 km (500 mi).

The next table presents the manufacturing energy, air emissions, and solid waste data associated with production of EPS insulation board (NREL 2005-present).

Table 4-14 EPS Foam Board Manufacturing Energy and Outputs

Manufacturing Energy	Unit	Quantity/kg foam	Unit	Quantity/lb foam
Electricity from grid	kWh	0.79	kWh	0.36
LPG in industrial boiler	L	2.00 E-03	gal	2.40 E-04
Natural gas in industrial boiler	m3	0.28	ft3	4.49
Outputs to air				
Carbon dioxide, fossil	kg	3.90E-03	lb	3.90E-03
Carbon monoxide, fossil	kg	2.10E-05	lb	2.10E-05
HFCs and HCFCs, unspecified	kg	1.10E-06	lb	1.10E-06
Nitrogen oxides	kg	4.60E-05	lb	4.60E-05
NMVOC, non-methane volatile organic compounds	kg	1.30E-04	lb	1.30E-04
Organic substances, unspecified	kg	1.10E-05	lb	1.10E-05
Particulates, unspecified	kg	3.20E-05	lb	3.20E-05
Pentane	kg	4.10E-02	lb	4.10E-02
Styrene	kg	2.20E-04	lb	2.20E-04
Sulfur oxides	kg	3.60E-07	lb	3.60E-07
Solid waste				
Solid waste to incineration with energy recovery	kg	1.70E-03	lb	1.70E-03
Solid waste to incineration without energy recovery	kg	1.80E-05	lb	1.80E-05
Solid waste to sanitary landfill	kg	2.10E-02	lb	2.10E-02

Regarding the pentane release, according to a document by the European Manufacturers of Expanded Polystyrene (EUMEPS, 2002), pentane is unstable and decomposes in the atmosphere into carbon dioxide and water within a few hours. Thus, all pentane, other than pentane captured and destroyed by onsite emission controls in place at many manufacturers, is assumed to be released at the manufacturing stage. This is consistent with Franklin (2009).

The solid waste is modeled as going to the landfill, incinerator, or WTE facility using ecoinvent data sets on waste management.

Transportation to the Building Site through End of Life

According to Franklin (2009), transportation of EPS to the building site is modeled to be an average of 483 km (300 mi) by heavy-duty diesel-fueled truck.

EPS is installed with an installation tape, but it is excluded since the tape makes up less than 1 % by weight so may be considered negligible. Pentane release at installation is considered negligible. Scrap EPS generated at installation is assumed to be 2 % of the total, consistent with other foam products in this category. While the product may be recyclable, it is modeled as being sent to a landfill 32 km (20 mi) from the building site. Data for the landfill come from waste management datasets in ecoinvent.

EPS foam board has a functional lifetime of over 40 years so no replacement is needed during the 40 year study period. How insulation in the buildings affects operating energy during the prototype buildings' use phase is addressed in other sections of this report.

While EPS foam is recyclable, it is assumed that it is disposed of in a landfill at end of life. End-of-life modeling includes transportation by heavy-duty diesel-fuel powered truck approximately 80 km (50 mi) to a C&D landfill. Insulation in a landfill is modeled based on ecoinvent end of life waste management process data.

Industry contact:

Walter A. Reiter, III, Esq., EPS Industry Alliance (2013)

XPS Foam Insulation

XPS foam insulation has been modeled for the exterior wall and roofing applications. Type IV XPS products used for roofing and commercial wall applications have a typical average density of 26.2 kg/m3 (1.63 lb/ft3). The following R-values are used in this study:[7]

- 2.5 cm (1 in): R = 5.0
- 5.1 cm (2 in): R = 10.6
- 7.6 cm (3 in): R = 16.2
- 10.2 cm (4 in): R = 22.0

Upstream Materials Production through Manufacturing

XPSA member companies provided representative industry average production data on XPS foam boards. XPSA represents the three largest producers in North America and accounts for over 95 % of XPS products produced and sold. The following table provides a 2010 representative average of the raw material and processing energy inputs and process outputs to produce one kg XPS foam board.

Table 4-15 XPS Foam Board Production Data

Inputs		Quantity/kg (Quantity/lb)
Blowing agents	HFC-134a kg (lb)	0.060 (0.060)
	HFC-152a kg (lb)	0.017 (0.017)
	CO_2 kg (lb)	0.012 (0.012)
Solid additives	PS resin kg (lb)	0.907 (0.907)
	Additives kg (lb)	0.018 (0.018)
Energy	Electricity kWh	1.00 (0.454)
Outputs		
Air	HFC-134a kg (lb)	0.0105 (0.0105)
	HFC-152a kg (lb)	0.0029 (0.0029)
Waste	Waste kg (lb)	1.0 E-4 (1.0 E-4)

The table presents the current representative blowing agent usage. It should be noted that HFCs began to replace HCFC-142b as the principal blowing agent in 2009, as the industry complied with U.S. EPA and Environment Canada ODS phase-out regulations requiring the XPS sector to exit HCFC use by the end of 2009. By 2010, all XPSA members had converted to non-HCFC blowing agents and have been using only HFC materials ever since.

The additives in the table include the flame retardant widely used in all XPS/EPS foams (hexabromocyclododecane (HBCD)) and colorants or dyes/pigments used to produce the characteristic color of each XPSA member's foam. Additives may also include a nucleation control agent, process lubricant, acid scavenger, or others.

---

[7] Extruded Polystyrene Foam Association (XPSA) website. http://www.xpsa.com. Values are based on a round-robin study in 2003 using the CAN/ULC S770-00 LTTR standard.

The blowing agent conversion/trim losses during manufacturing are assumed to be on average 17.5 % for North American XPS foam production (IPCC/TEAP, 2005, Table 7.7). All of the PS trim waste at the manufacturing plant is reused internally in the process. Only a very small amount of foam and other materials are occasionally sent off-site for disposal in a landfill.

Raw Materials Production

Data for polystyrene come from the U.S. LCI Database. Data for all three blowing agents and some of the additives come from ecoinvent. Data were not available for all of the additives; where data were not available proxy data were used.

All the raw materials are produced in the U.S., and most of the raw materials are centrally located as are the XPS manufacturer's largest plants. The estimated weighted average distance from the main suppliers to the majority of XPS manufacturing plants are 805 km (500 mi) for polystyrene, HFC-134a, flame retardant, and CO_2. HFC-152a and other additives are transported an average distance of 1609 km (1000 mi) to manufacturing plants. All but the blowing agents and polystyrene are transported by diesel truck; the blowing agents and polystyrene are transported by rail.

Transportation to the Building Site through End of Life

Transportation of the insulation to the building site is modeled as 563 km (350 mi), an average factoring in the various plants around the U.S. Transportation is by heavy-duty (combination) diesel-fueled truck.

Foam boards are installed with installation tape but tape is excluded since it is considered negligible. Scrap XPS foam board generated at installation is assumed to be 2 % of the total, consistent with other foam products in this category. While the product may be recyclable, it is modeled as being sent to a landfill 32 km (20 mi) from the building site. Data for the landfill come from waste management datasets in ecoinvent. Blowing agent escape during installation is insignificant. Minimal cutting to size on the jobsite is done and, even then, a sharp tool is typically used so that very few cells are opened.

XPS insulation has a functional lifetime of over 40 years so no replacement is needed during the 40 year study period. How insulation in the buildings affects operating energy during the prototype buildings' use phase is addressed in other sections of this report.

The diffusion of HFC-134a from XPS during use is 0.75 % +/-0.25 % per year (IPCC/TEAP, 2005, Table 7.7). The blowing agent emission loss during the use phase is complex and non-linear but can be represented for simplicity as a linear function after the first year. The rate is a function of the product thickness, properties (density, cell size, skins), blowing agent type(s) and transport properties (solubility, diffusion coefficient), and the installed application details (mean temperature, permeability of applied facings). The diffusion rate of HFC-152a is 15 % per year (IPCC/TEAP, 2005, Table 7.7).

Reuse of the foam is possible after building decommissioning, but the model assumes that at end of life the foam is disposed of in a landfill. End-of-life modeling includes transportation by heavy-duty diesel-fuel powered truck approximately 80 km (50 mi) to a C&D landfill. Insulation in a landfill is modeled based on ecoinvent end of life waste management process data.

For a typical NA building demolition followed by disposal in a landfill, it is reasonable to assume an initial blowing agent end of life loss of 20 % followed by annual losses of 1 % (UNEP/TEAP, 2005, Table 4.2).

Industry contact:
Comments and data provided by XPSA member companies and compiled by John Mutton, consultant to XPSA.

Polyisocyanurate (Polyiso) Foam Insulation
Polyiso foam insulation has been modeled for the exterior wall and roofing applications. The thermal resistance value for wall polyiso board is 2.9 $\frac{m^2*K}{W}$ per cm (6.5 $\frac{ft^2*K*h}{Btu}$ per inch), and for the roof it is 2.7 $\frac{m^2*K}{W}$ per cm (6.1 $\frac{ft^2*K*h}{Btu}$ per inch); the difference is due to the impermeable wall board facer vs. the permeable roof board facer. These R values are based on a 6-month accelerated aging test and were provided by representatives at Bayer MaterialScience.[8] The foam has a wet, or pre-yield, density of 29.2 kg/m^3 (1.82 lb/ft^3). The final product, which includes the weight of the facers, has a nominal density of 32.0 kg/m^3 (2.0 lb/ft^3).

Upstream Materials Production through Manufacturing

Upstream Materials Production

Cradle-to-gate data on production through manufacturing is based on a 2010 study performed for the Polyisocyanurate Insulation Manufacturers Association (PIMA) (Bayer MaterialScience, 2010). The scope of this study included collecting and compiling primarily 2007 production data from the six PIMA member companies and compiling it into an industry average polyiso insulation board. Process energy data came from 29 out of 31 polyiso plants in the U.S. and Canada, representing approximately 94 % of production in those geographies.

The chemicals to produce polyiso foam make up an "A" side (MDI) and a "B" side (polyester polyol with various additives such as catalysts, surfactants and flame retardants) plus a blowing agent (pentane). The table below presents the raw material inputs associated with polyiso foam production, provided on the basis of 2.54 cm (1 in) in Bayer MaterialScience (2010, Sec. 7.2).[9]

---

[8] Verbal communication with Bayer MaterialScience representatives, July 2013.
[9] The polyiso report presents corresponding values based on 6.2 cm (2.45 in) foam, the thickness used in the Polyiso LCA.

Table 4-16 Raw Material Inputs to Produce Polyiso Foam

Inputs:	% in foam (wt)	kg per 0.09 m2, 2.54 cm thick	lb per 1 ft2, 1 in thick
MDI	55.5	0.0382	0.0842
Polyester Polyol	31	0.0213	0.0470
TCPP	3.4	0.0023	0.0051
Catalyst K15	1.4	0.0010	0.0022
Catalyst PC46	0.16	1.38 E-04	0.0003
Catalyst PV	0.08	6.90 E-05	0.0002
Surfactant	0.63	5.51 E-04	0.0012
Pentane (blowing agent)	7.5	0.0052	0.0115
process water	0.1	0.0001	0.0002

The MDI comes from the U.S. LCI Database. Data for the polyester polyol comes from a 2010 Eco-Profile study of Aromatic Polyester Polyols (PE International, 2010). Data for Tris(2-chloroisopropyl)phosphate (TCPP) are U.S. data compiled from literature sources (PE International, 2011). Pentane data come from ecoinvent. No data were available to include the three catalysts or silicone surfactant; they total 2.3 % of the total input, so a total of 97.7 % of the inputs were included in the model.

There are two types of facers for polyiso foam insulation. Glass reinforced facer (GRF) is normally used in roofing applications and aluminum Kraft paper (foil) is normally used in wall applications. The GRF weighs 0.254 kg/m^2 (0.052 lb/ft^2) per layer, and 2 layers are used per foam board (totaling up to 13 cm or 5 in). The following table provides material and production energy for the GRF. The energy data come from an earlier version of the Polyiso LCA (Phelan, 2008), and materials and quantities have been compiled based on data from an MSDS (Atlas Roofing Corporation, 2008).

Table 4-17 GRF Process Inputs and Outputs

Energy and Material Inputs per 1 kg (per 2.2 lb):		
Corrugated cardboard and mixed fiber	kg (lb)	0.89 (1.96)[note 1]
Glass fiber	kg (lb)	0.10 (0.22)
Carbon black	kg (lb)	0.01 (0.02)
Electricity	MJ (Btu)	1.43 (1 355)
Nat gas	MJ (Btu)	9.08 (8 606)
Propane	MJ (Btu)	0.023 (21.8)

Outputs:		
Waste	kg (lb)	0.027 (0.06)

Note: the MSDS says 85 % comes from corrugated cardboard and mixed fiber, but 89 % are modeled to close the mass balance

According to Bayer MaterialScience (2010), 100 % of the fiber and glass (old corrugated cardboard, mixed paper, waste glass) are recycled materials, so their embodied contribution is burden-free. However, the energy to transport these materials to the facer manufacturing plant was included. Data for carbon black come from ecoinvent.

For the wall application, the foil facer raw materials include paper, aluminum foil, adhesives and coatings, and has a mass of 0.098 kg/m^2 (0.02 lb/ft^2) (Phelan, 2008). Data on material composition come from an MSDS; based on this limited data source, the facer is modeled as 77 % foil and 23 % Kraft (Atlas Roofing Corporation, 2012). Data for foil is modeled as 50/50 primary and secondary aluminum from the U.S. LCI Database, plus sheet rolling (ecoinvent). Data for Kraft paper come from ecoinvent.

Manufacturing

According to the Polyiso LCA, polyiso plants consume primarily electricity and natural gas used to operate the laminator and associated operations support equipment, such as thermal oxidizers, storage areas, packaging machines, raw material pumps, offices etc. A small amount of propane is used for fork lift trucks. The following table presents energy inputs and process outputs to produce 1 board-foot of foam, or 0.09 m^2 (1 ft^2) of 2.54 cm (1 in) thick polyiso foam.

Table 4-18 Energy Inputs and Process Outputs for 1 Board-Foot Polyiso Foam

Energy inputs	Unit	Quantity
Electricity	MJ (kWh)	0.0497 (0.0138)
Nat gas	MJ (Btu)	0.0913 (86.55)
Propane	kg (lb)	0.00015 (0.00031)

Outputs	Unit	Quantity
Pentane to air	kg (lb)	0.00013 (0.00030)
Waste scrap	board-foot	0.01

Based on review with polyiso plant manufacturers, approximately 2.5 % of the pentane added to the foam is lost to air during manufacturing. Depending on the plant and local regulatory requirements, pentane is emitted directly to the atmosphere or to a thermal oxidizer for combustion. Only 13 plants out of 31 use thermal oxidizers to combust the pentane emissions. Since the majority of polyiso plants in North America do not use thermal oxidizers, the pentane is modeled as going directly to atmosphere (Bayer MaterialScience, 2010).

Transportation and disposal of manufacturing waste scrap was modeled as going to an industrial landfill. It is assumed that a landfill for such non-hazardous waste is within 32 km (20 mi) of the polyiso plant.

Raw materials are transported to the manufacturing plant via diesel truck or rail. The following distances and modes of transport were modeled:

- MDI: 2 414 km (1 500 mi) by rail;
- Polyester polyol: 1 384 km (860 mi) by rail (90 %), 1 384 km (860 mi) by truck (10 %);
- Pentane: 2 414 km (1 500 mi) by truck;
- Remaining materials: 1 609 km (1 000 mi) by truck.

Transportation to the Building Site through End of Life

According to Bayer MaterialScience (2010), transportation to the building site is modeled as 400 km (250 mi) by heavy-duty (combination) diesel truck.

Installation tape is used but is excluded since it is considered negligible. Scrap polyiso generated at installation is assumed to be 2 % of the total, consistent with other foam products in this category. While the product may be recyclable, it is modeled as being sent to a landfill 32 km (20 mi) from the building site. Data for the landfill come from waste management datasets in ecoinvent. Pentane release at installation is negligible.

Polyiso insulation has a functional lifetime of over 40 years so no replacement is needed during the 40 year study period. How insulation in the buildings affects operating energy during the prototype buildings' use phase is addressed in other sections of this report.

Polyiso insulation is modeled as disposed of in a landfill at end of life. End-of-life modeling includes transportation by heavy-duty diesel-fuel powered truck approximately 80 km (50 mi) to a C&D landfill. Insulation in a landfill is modeled based on ecoinvent end of life waste management process data. According to Bayer MaterialScience (2010), 50 % of the total pentane in the product will have been released by end of life and 50 % remains in the product.

Roof Cover Boards

While a cover board may not be absolutely necessary, it is generally recommended to use over foam insulation as it creates a more durable roof system and better protects the underlying insulation board. As specified in RS Means for this study, perlite roof insulation was modeled in conjunction with the EPS; the function of which is to insulate in addition to being a protective covering. While these may legitimately still be used together, perlite is no longer commonly used in this application, according to industry experts. Gypsum/fiberglass products are more commonly used. DensDeck is one such product, and primary data were provided for this. Both perlite insulation and DensDeck are discussed below. Perlite is modeled only with the EPS while DensDeck is modeled with the XPS, and polyiso systems. DensDeck is modeled with EPS as an alternative scenario.

Perlite

Perlite roof insulation board is composed of expanded perlite particles, cellulose fibers, and binders. A one-inch board of perlite, with an R-value of 0.5 $\frac{m^2*K}{W}$ per cm (2.78 $\frac{ft^2*K*h}{Btu}$ per inch), is modeled with the EPS foam board. Perlite's density is 144 kg/m^3 (9 lb/ft^3) (GAF Materials Corporation, 2010). The table below presents the bill of materials in a perlite board. Data were compiled based on information from two MSDS's (GAF Materials Corporation, 2008; Johns Manville, 2010) so these estimated figures make more of a theoretical average.

Table 4-19 Perlite Board Bill of Materials

Materials	% (w/w)
Perlite (expanded)	68
Cellulose fiber	25
Asphalt, oxidized	5
Starch	2

Data for perlite come from ecoinvent. GAF, one perlite board manufacturer, states that minimum recycled content of its product is 25 %, so the cellulose fiber is modeled as post-consumer recycled paper. Data for the waste paper include transportation to a recycled paper processing facility and treatment (e.g., sorting). Data for asphalt come from the U.S. LCI Database. For starch, a potato starch data set from a Danish food LCA database is used (LCAFood database).

The raw materials are modeled as transported to the manufacturing plant via combination diesel truck. Materials are assumed to be sourced domestically; it is assumed the average transportation distance is 805 km (500 mi).

Generally, perlite board may be made by expanding the perlite, adding a tacky resin emulsion to the perlite, drying the mixture and forming it into a board-like product (Nath, 1982). The ecoinvent data set for perlite includes processing of perlite, so this is included in the model. Additionally, some of data for manufacturing particle board were used as proxy manufacturing data since the processes have similarities (Wilson, 2008).

Transportation of the final product to the building site is modeled as 1 609 km (1 000 mi) by heavy-duty diesel-fueled truck.

At end of life, it is assumed that it is disposed of in a landfill. End-of-life modeling includes transportation by heavy-duty diesel-fuel powered truck approximately 80 km (50 mi) to a C&D landfill. The product in a landfill is modeled based on ecoinvent end of life waste management process data.

Gypsum/Fiberglass Coverboard

DensDeck is a fiberglass mat-faced, treated gypsum core panel. A 0.64 cm (0.25 in) DensDeck product has been modeled in all the foam insulation board systems. The main bill of materials for DensDeck is presented in the table below. The full bill of materials was provided by Georgia Pacific for use in this specific project but these data are not presented to protect confidentiality. The table therefore contains data from a Georgia Pacific MSDS (2009).

Table 4-20 DensDeck Bill of Materials

Materials	%	Comments
Gypsum 60 % to 100 %	95	* The MSDS says the gypsum contains naturally occurring silica crystalline (quartz), in the amount of 0.1 % to 1 % of total product
Continuous filament GF: 1 % to 5 %	5	

Both gypsum and glass fiber are modeled based on ecoinvent data sets. Materials to manufacture DensDeck come from domestic and foreign sources. For the domestic transportation, a distance of 805 km (500 mi) by heavy-duty diesel truck was assumed. For foreign sources, an average of 2 414 km (1 500 mi) by barge has been assumed. Actual distances were provided and modeled for this project but these data are not presented to protect confidentiality.

Transportation of the final product to the building sites is done by truck (95 %) and rail (5 %). Due to the several Georgia Pacific plants producing DensDeck, an average of 402 km (250 mi) by truck and 1 127 km (700 mi) by rail has been modeled.

At end of life, it is assumed that it is disposed of in a landfill. End-of-life modeling includes transportation by heavy-duty diesel-fuel powered truck approximately 80 km (50 mi) to a C&D landfill. Insulation in a landfill is modeled based on ecoinvent end of life waste management process data.

4.5.3 Windows

Introduction

BIRDS evaluates aluminum single- and double-pane windows and double- and triple-pane windows with thermal breaks, as shown in the table below. Not shown in the table are fiberglass and wood-PVC composite frame operable windows which are used for some of the high energy efficient applications in BIRDS. Also not included in the table are low-emissivity (low-E) coating, tint, and reflective coatings, which were modeled as part of the window systems in order to meet the windows' U-factor, solar heat gain coefficient (SHGC), and visible transmittance (VT) performance requirements for each particular climate zone and code edition.

Table 4-21 Window Types in BIRDS

Type of Building	Window Type	Frame Type(s) and # Panes	Area	Window to Wall Ratio (WWR)
4-story & 6-story apt buildings, 4-story & 6-story dorms	**Operable sliding windows**	Al frame - single pane Al frame - double pane Thermo-break Al frame - double pane	152 cm (5 ft) by 91 cm (3 ft)	4-story apt building = 12 %; 6-story apt building = 14 %; 4-story & 6-story dorms = 20 %
Elementary school, 8-story office building, 3-story office building, high school	**Punched opening, fixed windows**	Al frame - single pane Al frame - double pane Thermo-break Al frame - double pane	135 cm (4 ft 5 in) by 160 cm (5 ft 3 in)	Elementary & high schools = 25 %; 8-story & 3-story office = 20 %
Retail store, restaurant	**Fixed storefront**	Al frame - single pane Al frame - double pane Thermo-break Al frame - double pane	Typical/common storefront fixed window area	Retail store = 10 %; Restaurant = 30 %
16-story office building, 15-story hotel	**Curtain wall**	Al frame - single pane Al frame - double pane Thermo-break Al frame - double pane Thermo-break Al frame - triple pane	Common lengths of sticks, components, pieces for a curtain wall	Office building & hotel = 100 %

It is acknowledged that there are multiple window assembly options (combination of frame material, glass in-fill, and operability) for any building type in any climate zone; multiple window assembly options can be nearly identical in performance. The window assembly types presented in these tables are only one of many options available. As such, the particular window assembly combinations presented in this documentation are not endorsed or preferred over any other type of window assembly for the respective buildings in which they are used in BIRDS.

It should also be noted that window to wall ratios (WWRs) vary greatly for many different reasons. Those selected for use in BIRDS are based solely on information provided by the RS Means *Costworks*, and may not be representative of WWRs for actual buildings of similar types. The next iteration of BIRDS will include additional prototypes such as the DOE Benchmark Buildings, which are expected to provide more representative WWRs. Additionally, industry members will be able to advise on the WWR aspect of the prototype buildings.

BIRDS environmental performance data for the windows category was provided on a per-window basis for the operable sliding windows and fixed windows, and per 0.09 m^2 (1 ft^2) of typical or common size of fixed storefront and curtain wall windows. The flow diagram below presents the general system boundaries for the window category as it is modeled for BIRDS.

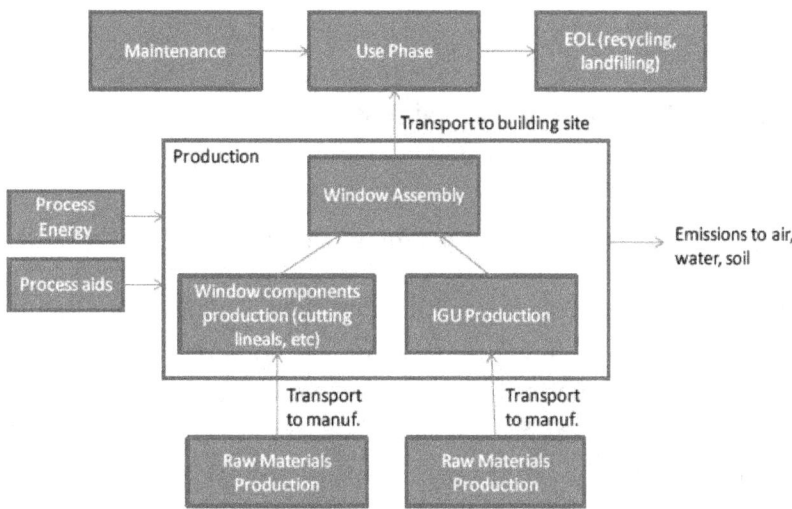

Figure 4-7 Windows System Boundaries

Aluminum Operable Sliding Window
Material take-off data for 2010 production were provided for an aluminum operable sliding window of dimensions 152.4 cm (5 ft) by 91.4 cm (3 ft). The following table provides general data on the dimensions and main components of a single- and double-pane window, and a double-pane window with a thermal break frame. A thermal break is a material that separates the interior and exterior of an aluminum (or other metal) frame. The low thermal conductivity of the material reduces heat transfer, making the metal window energy efficient. Due to the proprietary nature of the data, the full details of all of the materials have been removed.

Table 4-22 Characteristics and Components of the Aluminum Operable Window

Description	Units	Single pane	Double pane	Double pane w/ thermo-br
		Quantity		
Weight of the finished sealed unit	kg (lb)	33 (72)	47 (103)	48 (105)
Dimension of the window	m2 (ft2)	1.4 (15)	1.4 (15)	1.4 (15)
Depth of the finished sealed unit	cm (in)	8.3 (3.25)	8.3 (3.25)	8.3 (3.25)
Extruded aluminum (frame)	kg (lb)	16 (35)	15 (33)	15 (33)
IGU	kg (lb)	15 (32)	29 (65)	29 (65)
Other components	kg (lb)	2.10 (4.64)	2.11 (4.65)	2.95 (6.50)

Greater than 99 % of the mass of materials in the window were included and modeled for this window system. The aluminum frame is modeled as 50 % primary and 50 % secondary extruded aluminum, with the LCI data coming from the U.S. LCI Database. "Other components" include clips, weatherstrip, gaskets, hardware, sealant, and thermobreak materials. Clips, weatherstrip, and gaskets are modeled as PVC and rubber parts. Hardware is custom ordered so may vary with the window. For BIRDS, a mix of 50 % stainless steel and 50 % bronze has been modeled. The sealant is assumed to be silicone-based, and thermal break materials include glass-reinforced polyamide and PUR resin. U.S. LCI Database provided the production data for PVC while the ecoinvent database provided the data for the remaining materials.

Data for the insulated glass units (IGU) are compiled from two sources: the commercial window manufacturer and Salazar (2007, Table 5.4). These two data sources have been averaged into one "hypothetical" IGU with some categories having combinations of materials in order to increase representation of an IGU in the marketplace, which has an array of materials that can be used for various parts of the IGU. The bill of materials on the basis of 0.09 m^2 (1 ft^2) of the IGU is as follows:

Table 4-23 Insulated Glass Unit Materials

IGU category	Material	kg / m^2	lb / ft^2
Glass (2 panes)	Glass (0.25")	14.1	2.9
Gas filler	Argon	2.01E-02	4.12E-03
Spacer	50/50 Stainless St/Alum	1.86E-01	3.81E-02
Primary sealant	Polyisobutylene	7.70E-03	1.58E-03
Secondary sealant	50/50 Polysulfide/PUR	1.71E-01	3.51E-02
Desiccant	Silicone	8.98E-02	1.84E-02

The argon was calculated based on the volume between the panes of glass separated by 1.27 cm (0.5 in) and an argon density of 1.65 kg/m^3 (0.103 lb/ft^3). Salazar (2007) reported an escape of argon when the cavity is flushed prior to being sealed – approximately 95 % more than the quantity in the finished unit. The manufacturer did not provide data on material losses. Since it is unknown if this quantity of escaped argon is high or low, a 50 % escape is assumed for this model. The spacer, which separates the two panes of glass, can be made of an array of materials, including aluminum, stainless steel, and structural foam. Since no data were provided on an IGU with structural foam, a 50/50 assumption of aluminum and stainless steel was used. Salazar (2007) reported a loss of approximately 10 % of the spacer; this has been factored in to the model. However, for both materials, the loss is recycled and not actually waste. The inner primary sealant, commonly polyisobutylene (PIB), is used to prevent leakage of the argon gas as well as penetration of water vapor into the space between the panes. An assumption of 50/50 polysulphide polymer and polyurethane as secondary sealants were used. The desiccant in the spacer, assumed to be a silicone product, is used to absorb moisture. Salazar (2007) reported a loss of approximately 0.7 % and this has been factored into the model. Losses discussed in this paragraph are not included in the table above.

Note that for windows specified without argon filling, the IGU bill of materials is the same as the table above without the argon filling. Also, coatings and tints, when applicable, have been included in the model but are not included in the table.

The data for float glass come from ecoinvent, based on early 2000's European processes and technologies. Due to lack of available data on U.S. float glass production, the older European data were used. Processes in the data set include melting, cullet addition, forming (on a float bath), annealing by cooling in an oven (lehr), cutting of the glass, and storage. While this data set may not be representative of current U.S. production, it has been customized using U.S. energy and transportation data sets. Also, some transportation impacts have been removed, including transport between manufacturing plants and coating facilities, which, according to U.S. windows industry representatives, exists for European operations but not for U.S. operations. The next version of BIRDS is expected to have more improved data on glass production.

The argon, stainless steel, polysulfide, and desiccant data come from ecoinvent. PIB come from ecoinvent; synthetic rubber is used as a general proxy for PIB. Aluminum and PUR data come from the U.S. LCI Database.

No data were provided on the supplier distances to the manufacturer, so it is assumed that the materials to the manufacturing plant are transported an average of 600 km (373 mi) by heavy-duty diesel-fueled truck.

Aluminum Punched Opening, Fixed Window Production
Material take-off data for 2010 production were provided for an aluminum punched opening, fixed window of dimensions 135 cm (4 ft 5 in) by 160 cm (5 ft 3 in). The following table

provides general data on the dimensions and main components of a single- and double-pane window, and a double-pane window with a thermal break frame. Due to the proprietary nature of the data, the full details of all of the materials have been removed.

Table 4-24 Characteristics and Components of the Aluminum Fixed Window

Description	Units	Single pane	Double pane	Double pane w/ thermo-br
			Quantity	
Weight of the finished sealed unit	kg (lb)	39 (85)	66 (145)	67 (148)
Dimension of the window	m^2 (ft^2)	1.4 (15)	1.4 (15)	1.4 (15)
Depth of the finished sealed unit	cm (in)	5.1 (2.0)	5.1 (2.0)	5.1 (2.0)
Extruded aluminum (frame)	kg (lb)	10 (23)	10 (22)	11 (24)
IGU	kg (lb)	27 (60)	55 (122)	55 (122)
Other components	kg (lb)	0.93 (2.05)	0.78 (1.72)	0.78 (1.72)

Greater than 99 % of the mass of materials in the window were included and modeled for this window system. The aluminum frame is modeled as 50/50 primary and secondary extruded aluminum, with the LCI data coming from the U.S. LCI Database. "Other components" include clips, gaskets, sealant, and thermal break materials. Clips and gaskets are modeled as PVC and rubber parts. The sealant is assumed to be silicone-based, and thermal break materials include glass-reinforced polyamide and PUR resin. U.S. LCI Database provided the production data for PVC while the ecoinvent database provided the data for the remaining materials. No data were provided on the supplier distances to the manufacturer, so it is assumed that the materials to the manufacturing plant are transported an average of 600 km (373 mi) by heavy-duty diesel-fueled truck.

The IGU for the fixed, punched opening window is the same as for the operable sliding window. Please refer to that section above.

Aluminum Fixed Storefront Window Production
Material take-off data for 2010 production were provided for an aluminum fixed storefront window of dimensions 203 cm (80 in) by 203 cm (80 in). The following table provides general data on the dimensions and main components of a single- and double-pane window, and a double-pane window with a thermal break frame. Due to the proprietary nature of the data, the full details of all of the materials have been removed.

Table 4-25 Characteristics and Components of the Aluminum Storefront Window

Description	Units	Single pane	Double pane	Double pane w/ thermal-br
			Quantity	
Weight of the finished sealed unit	kg (lb)	78 (173)	141 (310)	133 (293)
Dimension of the window	m² (ft²)	4.1 (44)	4.1 (44)	4.1 (44)
Depth of the finished sealed unit	cm (in)	11.4 (4.5)	11.4 (4.5)	11.4 (4.5)
Extruded aluminum (frame)	kg (lb)	24 (52)	30 (66)	21 (46)
IGU	kg (lb)	54 (118)	109 (240)	109 (240)
Other components	kg (lb)	1.8 (4.0)	1.9 (4.3)	3.2 (7.2)

Greater than 99 % of the mass of materials in the window were included and modeled for this window system. The aluminum frame is modeled as 50/50 primary and secondary extruded aluminum, with the LCI data coming from the U.S. LCI Database. "Other components" include clips, gaskets, sealant, and thermal break materials. Clips and gaskets are modeled as PVC and rubber parts. Thermal break materials include glass-reinforced polyamide and PUR resin. U.S. LCI Database provided the production data for PVC while the ecoinvent database provided the data for the remaining materials. No data were provided on the supplier distances to the manufacturer, so it is assumed that all materials to the manufacturing plant are transported 600 km (373 mi) by heavy-duty diesel-fueled truck.

The IGU for the fixed storefront window is the same as for the operable sliding window. Please refer to that section above.

Aluminum Curtain Wall Window Production

Material take-off data for 2010 production were provided for an aluminum curtain wall window of dimensions 203 cm (80 in) by 203 cm (80 in). The following table provides general data on the dimensions and main components of a single- and double-pane window, a double-pane window with a thermal break frame, and a triple-pane window with a thermal break frame. Due to the proprietary nature of the data, the full details of all of the materials have been removed.

Table 4-26 Characteristics and Components of the Aluminum Storefront Window

Description	Units	Quantity			
		Single pane	Double pane	Double pane w/ thermal-br	Triple pane w/ thermal-br
Weight of the finished sealed unit	kg (lb)	105 (232)	157 (346)	158 (348)	213 (469)
Dimension of the window	m² (ft²)	4.1 (44)	4.1 (44)	4.1 (44)	4.1 (44)
Depth of the finished sealed unit	cm (in)	17.8 (7.0)	17.8 (7.0)	17.8 (7.0)	17.8 (7.0)
Extruded aluminum (frame)	kg (lb)	49 (108)	47 (103)	49 (107)	51 (112)
IGU	kg (lb)	53 (116)	107 (236)	107 (236)	159 (350)
Other components	kg (lb)	3.4 (7.5)	3.5 (7.8)	2.3 (5.1)	3.1 (6.7)

Greater than 99 % of the mass of materials in the window were included and modeled for this window system. The aluminum is modeled as primary extruded aluminum, with the LCI data coming from the U.S. LCI Database. "Other components" include clips, gaskets, sealant, and thermal break materials. Clips and gaskets are modeled as PVC and rubber parts. Thermal break materials include glass-reinforced polyamide and PUR resin. U.S. LCI Database provided the production data for PVC while the ecoinvent database provided the data for the remaining materials. No data were provided on the supplier distances to the manufacturer, so it is assumed that all materials to the manufacturing plant are transported 600 km (373 mi) by heavy-duty diesel-fueled truck.

The IGU for the curtain wall is essentially the same as for the operable sliding window. Please refer to that section above.

Fiberglass Frame Operable Window

Salazar (2007, Sec. 5.1.3) provided the bill of materials and production details for a fiberglass casement window with dimensions 60 cm (1.97 ft) by 120 cm (3.94 ft). Data are from 2004 and came from a manufacturer who produces fiberglass pultrusions and pultrusion machinery in addition to windows. The finished frame includes the fiberglass lineals, polyester bracing in the corners, and a PVC glazing stop. The following table provides the bill of materials in the window (Salazar, 2007, Table 5.8). Since these materials were based on a perimeter of 3.6 m (11.7 ft), the quantities of the lineal materials were scaled up a factor of 1.4 to meet the 4.9 m (16.0 ft) perimeter used in the BIRDS tool.

Table 4-27 Bill of Materials of the Fiberglass Operable Window

Material / component	Window 0.6 m (2.0 ft) x 1.2 m (3.9 ft) kg	(lb)	Window 0.9 m (3 ft) x 1.5 m (5 ft) kg	(lb)
Textile Glass (Roving)*	2.80	(6.17)	3.82	(8.43)
Textile Glass (Mat)*	1.40	(3.09)	1.91	(4.21)
Polystyrene Resin*	2.10	(4.63)	2.87	(6.32)
Calcium Carbonate*	0.70	(1.54)	0.96	(2.11)
PVC	0.497	(1.10)	0.497	(1.10)
Polyester	0.37	(0.82)	0.37	(0.82)
Steel (Operator)	1.88	(4.14)	1.88	(4.14)
Steel (Fasteners)	0.088	(0.19)	0.088	(0.19)

* Data for these materials are approximate to protect proprietary manufacturer data

The fiberglass pultrusion lineals are assembled with a purchased sealed unit measuring 0.48 m^2 (5.2 ft^2).

Textile glass is modeled using ecoinvent data for glass fiber. The glass mat is modeled as it is in Salazar (2007), i.e., containing 100 % recycled A-glass from light bulbs. Polystyrene resin is based on the general purpose polystyrene data set from the U.S. LCI Database. U.S. LCI Database also provided the data for the PVC resin and CaCO3. Polyester resin comes from ecoinvent. Steel fasteners and the operator come from World Steel Association (2011), with steel profiles customized to U.S. using the U.S. electricity grid.

The IGU is modeled the same as the other IGUs in the window category (described above).

It is assumed that materials are transported to the manufacturing plant an average of 600 km (373 mi) by heavy-duty diesel-fueled truck.

Wood- PVC Composite Frame Operable Window

Limited data were available for a wood-PVC composite window frame, so some data were extrapolated from Salazar (2007)'s aluminum clad wood window (Sections 5.1.1, 5.2.1). Assuming that the wood-PVC composite window has the same volume of material in the frame, the volume of wood in the Salazar window was used to estimate the quantity of composite material in the frame. Other components from the wood window were used in this model, including the fasteners, hardware, and weatherstrip. Since these materials were based on a perimeter of 357 cm (11.7 ft), the quantities relating to the perimeter of the frame were scaled up a factor of 1.4 to meet the 488 cm (16 ft) perimeter used in the BIRDS tool. The following table presents the bill of materials in the window.

Table 4-28 Bill of Materials of the Wood-PVC Composite Operable Window

Material / component	Window 0.6 m (2.0 ft) x 1.2 m (3.9 ft)		Window 0.9 m (3 ft) x 1.5 m (5 ft)	
	kg	(lb)	kg	(lb)
Wood-PVC Composite: 0.07 m3*	63.0	(138.9)	86.0	(189.6)
Paint	0.30	(0.66)	0.41	(0.90)
Polypropylene	0.12	(0.26)	0.16	(0.36)
Thermoplastic Elastomer	0.09	(0.20)	0.12	(0.27)
Steel (Operator)	1.79	(3.95)	1.79	(3.95)
Steel (Fasteners)	0.16	(0.35)	0.16	(0.35)

*Composite density = 900 kg/m3 (56.2 lb/ft3) (Flakeboard, www.flakeboard.com)

Composite material is modeled as being made up of 40 % fine wood fiber (flour or particles) and 60 % thermoplastic polymer by weight[10] (assuming to be PVC in this case). The composite is produced by mixing the wood fiber with the heated resin. It is assumed that the wood fiber is pre-consumer recycled content. One provider of wood-PVC composite windows stated that "some" of the polymer has been reclaimed so 10 % was modeled as pre-consumer reclaimed PVC. Data for virgin PVC come from the U.S. LCI Database. The formation of the composite lineals is assumed to be done by extrusion, and this is based on ecoinvent data.

Paint is modeled as alkyd paint and comes from ecoinvent. Data for polypropylene come from the U.S. LCI Database. Data for the thermoplastic elastomer, as ethylene propylene diene monomer (EPDM) rubber, come from ecoinvent. Steel fasteners and the operator are based on cold rolled steel data from World Steel Association (2011).

The IGU is modeled the same as the other IGUs in the window category (described above).

It is assumed that materials are transported to the manufacturing plant an average of 600 km (373 mi) by heavy-duty diesel-fueled truck.

Window Manufacturing
<u>Aluminum Windows</u>

The large commercial window manufacturer provided electricity, natural gas energy, and net water use data on the basis of one operable or fixed window, storefront window, and curtainwall window. For confidentiality purposes, these data are not shared in this documentation, however, they have been included in the models for all of the window types.

---

[10] Website information on Fibrex Material. Found at:
http://www.andersenwindows.com/planning/articles/fibrex-material/.

Fiberglass Windows

Manufacturing energy was provided by the fiberglass manufacturer, but the data were limited due to unavailable data on total windows produced in the given year as well as the broad variation of business practices of this manufacturer. Nonetheless, an allocation was made for the pultrusion energy in Salazar (2007), so those data were used for pultrusion: 20 MJ (18 956 Btu) natural gas and 24.2 MJ (22 937 Btu) electricity per window (Table 5.15). Energy data used for the aluminum windows were applied for the remaining processes, i.e., window assembly and IGU production. These energy data are not disclosed to protect confidential data.

According to Salazar (2007), the heat curing during pultrusion liberates approximately 5.5 % of the styrene into the air, or 0.112 kg (0.247 lb). Waste is also generated and included in the model: waste comes from the cutting of lineals (0.486 kg (1.07 lb)) and glazing stops (0.008 kg (0.018 lb)) to length. This waste is modeled as going to a landfill.

Wood-PVC Composite Windows

Total facility energy was provided by the wood window manufacturer, and the Salazar study allocated these totals to four "departments": lineal production, assembly, overhead, and sealed units. As mentioned above, the lineals were produced by way of extrusion, and this is accounted for in the production of upstream materials. The same energy data used to produce the IGUs for the aluminum windows were used here. Assembly energy, which included cutting lineals to length, installing hardware, and placing the sealed unit in the frame, plus overhead energy, were taken from the Salazar study. Manufacturing energy was as follows:

Table 4-29 Wood-PVC Composite Window Manufacturing Energy

Facility Aspect	Natural Gas MJ (Btu)[11]	Electricity MJ (Btu)
Window Assembly	90.4 (85 683)	61.5 (58 291)
Overhead	43.0 (40 756)	29.2 (27 676)

It is assumed that 5 % to 10 % (average 7.5 %) of the lineal is wasted during cutting to size, but this material can be re-melted and re-extruded so is not considered as a waste.

Coatings

Low-emissivity (low-E) coatings, reflective coatings, and tinted windows have been included in the windows modeling to meet different performance characteristics of the windows. Coatings are used to improve the insulation properties of the glass by reflecting visible light and infrared

---

[11] The manufacturer in Salazar (2007) used wood as a significant source of energy. Since a manufacturer of composite windows may not necessarily produce wood windows (nor have wood waste as a readily available energy source), the wood and natural gas energy quantities were combined into only natural gas.

radiation. Low-E coating is modeled using the coating details of ecoinvent's "flat glass, coated" data set as a starting point. The technology used at this plant is based on a cathodic sputtering technology which involves depositing thin silver and other layer(s) on the glass. According to the ecoinvent documentation, the raw materials used for sputtering are bismuth, silver and nickel-chrome. The quantity of 1.19 E-4 kg (2.62 E-4 lb) metals per kg was divided into three to account for 1/3 nickel, 1/3 chromium, and 1/3 silver. It is acknowledged that these data are approximate.

According to the Windows for High Performance Commercial Buildings website, reflective coatings usually consist of thin metallic or metal oxide layers and come in various metallic colors including silver, gold, and bronze.[12] Due to lack of available other data, the reflective coating data use the same quantity of coating as low-E but apply silver, gold and bronze data sets to account for a range of reflective coatings.

Tint is obtained by adding small amounts of metal oxides during glass manufacturing, coloring the glass uniformly. For BIRDS, iron oxide has been assumed to be the mineral additive for the tint, and it is modeled as applied at an assumed rate of 0.1 % of the weight of glass.

Transportation to the Building Site, Use and Maintenance
Transportation of the window to the building site is modeled an average of 805 km (500 mi) by heavy-duty (combination) diesel fuel-powered truck.

Installing windows is primarily a manual process; no energy or emissions are included in the model for this. Windows come to the building fully assembled and custom-ordered to fit so there is generally no installation waste.

The commercial windows are modeled as having a functional lifetime of over 40 years so no replacement is needed during the 40 year study period. Maintenance in the model includes weatherstripping and sealing (discussed below). All operational energy-related aspects of the window are addressed in other sections of this report.

Weatherstripping

Weatherstrip is modeled as a thermoplastic elastomer. Data for the thermoplastic, as ethylene propylene diene monomer (EPDM) rubber, come from ecoinvent (as synthetic rubber). For BIRDS, an EPDM weatherstrip has been modeled in the amount of 0.0064 kg per 0.3 m (0.014 lb per ft).

Weatherstrip is assumed to perform at its optimal level an average of 7.5 years (Vigener, 2012) so is modeled as replaced every 7.5 years.

---

[12] Site developed jointly by the University of Minnesota and Lawrence Berkeley National Laboratory. Found at: http://www.commercialwindows.org/reflective.php.

Different perimeter sealants can be used for different applications. For BIRDS, a silicone sealant has been modeled in the amount of 0.018 kg per 0.3 m (0.04 lb per ft). This is based on the average amount of sealant needed for perimeter depth/width gaps of 0.64 cm/0.64 cm (0.25 in/0.25 in) and 0.64 cm/1.27 cm (0.25 in/0.5 in).[13] Data for the sealant comes from ecoinvent, primarily its silicone product. The sealant is modeled as being replaced every 15 years (Vigener, 2012).

Other maintenance, such as glass and/or window frame cleaning, frame repainting or recoating, hardware adjustment or replacement, etc., are not included in the analysis.

End of Life

The aluminum frame portion of the window is modeled as recycled at end of life, and the 0-100 recycling methodology has been applied. For this, system expansion is applied; the production of the same amount of virgin aluminum that is in the frame is subtracted out of the system, crediting the system with an avoided burden based on the reduced requirement for virgin material production in the next life cycle. Likewise, recycled content in the aluminum adds some of the burden to the product system in order to share the burden with the previous life cycle.[14]

The remaining parts of the window, including the IGU, are disposed of in a landfill. Fiberglass and composite frames are modeled as disposed of in a landfill. End-of-life modeling includes transportation by diesel-fuel powered truck approximately 80 km (50 mi) to a C&D landfill or to recycling. The portions of the window going to landfill are modeled based on ecoinvent end of life waste management process data.

4.5.4   HVAC Systems

The HVAC systems for each building type in BIRDS were based on the default prototype specifications provided by the RS Means *Square Foot Cost Estimator* (*SFCE*). BIRDS evaluates four main categories of HVAC systems: fin tube radiation, boilers, packaged chillers, and rooftop air conditioners. Data presented in Section 3 lay out the operational characteristics and applications in the prototype buildings. The subsections below describe the life cycle modeling of the equipment.

BIRDS environmental performance data for the HVAC equipment evaluated was provided on a per-unit basis. The flow diagram below presents the general system boundaries for the HVAC equipment as it is modeled for BIRDS.

---

[13] Amount calculated on an on-line sealant usage calculator, found at http://www.tremcosealants.com/technical-resources/calculators/sealant-calculator.aspx.

[14] For more information on the approach to modeling metals at end of life, see Atherton (2006) and World Steel Association (2011).

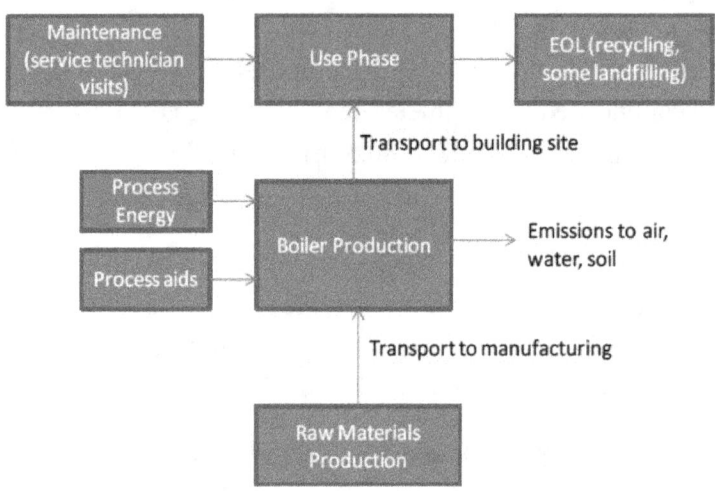

Figure 4-8 HVAC System Boundaries – Boiler Example

Boilers

Boilers provide the steam or hot water to various parts of the buildings. The following types of boilers were evaluated:

Table 4-30 Boilers Included in BIRDS

Type of Boiler, construction	Gross output in kW (MBH[15])	Notes
Cast iron, natural gas- or propane-fired	Steam, 70 (240)	Includes standard controls and insulated jacket
Cast iron, natural gas- or propane-fired	Steam, 82 (280)	Includes standard controls and insulated jacket
Cast iron, natural gas- or propane-fired	Steam, 48 (163)	Includes standard controls and insulated jacket
Cast iron, natural gas- or propane-fired	Steam, 448 (1 530)	Includes standard controls and insulated jacket
Cast iron, natural gas- or propane-fired	Steam, 1 958 (6 680)	Includes standard controls and insulated jacket
Cast iron, gas- and oil-fired	Hot water, 1 758 (6 000)	

Upstream Materials Production through Manufacturing

Data were compiled from publicly-available manufacturer-specific literature data (table below). These specific models were chosen since their literature provided adequate data to compile a rough estimate of the material quantities.

---

[15] MBH = 1 000 Btu/hr.

Table 4-31 Boiler Models

Output, kW (MBH)	Manufacturer & model	Data Source for dimensions, weight*
Steam, 70 (240) Steam, 82 (280) Steam, 48 (163)	Weil McClain, EGH 65-75 Weil McClain, EGH 75-85 Weil McClain, EGH 50-55	Weil McClain EG-PEG-EGH Gas Fired Boilers Series 4 – Boiler Manual, Part No. 550-110-639/0706, Table 15 & p. 34 diagrams.
Steam, 448 (1 530)	Weil McClain, LGB 12-13	Weil McClain LGB Gas Fired Boiler – Boiler Manual, Part Number 550-141-186/0703, Sections 12, 14.
Steam, 1 958 (6 680) Water, 1 758 (6 000)	Weil McClain, (94)H2494 Weil McClain, (94)H2194-2294	Weil-McLain 94 Water and steam boilers — for Gas, Light Oil, & Gas/Light Oil-Fired Burners – Boiler Manual, Part No. 550-110-275/0810, p.42 & Fig 5.
*Weil-McLain boiler manuals retrieved at: http://www.weil-mclain.com/en/our-products/boilers/commercial-boilers.aspx		

As a starting point for the bill of materials, data on the cast iron content of a 938 kW (3 200 MBH) Veissmann water boiler which has a cast iron heat exchanger was assumed (Veissmann Vitorond, July 2010),[16] and this content was approximately 96 %. The assumption was then made that this percentage was the same for other boilers with cast iron heat exchangers, and a representative from Weil McClain confirmed that it is "roughly 95 %".[17] Thus, cast iron was assumed for 94 % for smaller output boilers and 96 % for larger ones. The remaining 4 % to 6 % of the total weights for the steel jacket and volume of insulation were calculated based on dimensions of the units and an insulation thickness of 1.3 cm (0.5 in) of fiberglass batt (thickness confirmed by Weil McClain). The weight of fiberglass was calculated using a density of 22.6 kg/m^3 (1.41 lb/ft^3) (see insulation section). The remaining weight was assumed to be cold rolled steel and a small amount of copper in the controls. The following table provides the bill of materials for each boiler based on the above assumptions and boiler weight data from literature.

---

[16] For example: the Viessmann model Vitorond VD2 1080.
[17] Correspondence with Weil McClain, February 2012.

Table 4-32 Boiler Bill of Materials

Output, kW (MBH)	Total kg (lb)	Cast Iron kg (lb)	Steel kg (lb)	Insulation kg (lb)	Copper kg (lb)
Steam, 70 (240)	320 (705)	301 (663)	18 (40)	1.0 (2.3)	0.11 (0.25)
Steam, 82 (280)	358 (789)	337 (742)	20 (45)	1.1 (2.5)	0.11 (0.25)
Steam, 48 (163)	265 (585)	249 (550)	15 (33)	0.9 (2.0)	0.11 (0.25)
Steam, 448 (1 530)	2 064 (4 551)	1 981 (4 368)	79 (175)	3.1 (6.9)	0.36 (0.8)
Steam, 1 958 (6 680)	10 161 (22 400)	9 754 (21 503)	397 (876)	9.1 (20.0)	0.45 (1.0)
Water, 1 758 (6 000)	9 219 (20 324)	8 850 (19 510)	361 (795)	8.4 (18.5)	0.45 (1.0)

Ecoinvent data were used for the cast iron. Data for the cold rolled steel come from World Steel Association (2011), and data for copper come from ICA (2012). The fiberglass insulation was modeled as described in the insulation section.

Raw materials are modeled as transported to the manufacturing plant via diesel truck an assumed average distance of 805 km (500 mi).

No manufacturing data were available for commercial boilers, so as proxy, an ecoinvent data set for a 100 kW (341 MBH) oil boiler was used (ecoinvent Report No. 5, 2007). The ecoinvent data summary describes data for the production energy to be estimated and coming from a 1998 environmental report. In general, the data, being estimated and of older vintage, are considered to be not representative. Nonetheless, the use of these data was considered to be better than no data at all.

Table 4-33 Boiler Manufacturing Data

Energy source	Quantity for 580 kg boiler 100 kW
Electricity, medium voltage MJ (Btu)	1 195 (1 132 831)
Natural gas in industrial furnace MJ (Btu)	1 920 (1 819 809)
Light fuel oil in industrial furnace MJ (Btu)	1 010 (957 295)
Tap water liter (gal)	741 (196)

Since manufacturing larger or smaller units requires more or less energy, respectively, these data were normalized up or down based on the total weights of the units.

Transportation to the Building Site through End of Life

Transportation of the equipment to the building site is modeled to be an assumed average of 644 km (400 mi) by heavy-duty diesel fuel-powered truck.

It is assumed that a qualified service technician comes to the building site to check and/or service the unit one time per year to ensure optimal performance and lifetime. It is assumed that the qualified technician is within a 24 km (15 mi) service radius. The 24 km (15 mi), driven in a gasoline-powered van, is allocated amongst other service visits for that technician, assuming that the same technician is making more than one service call during that trip. Assuming the technician makes 5 service calls in one day, one-fifth of the impacts from driving 24 km (15 mi) are allocated to the product, or 4.8 km (3 mi). Data for a van come from ecoinvent. Unplanned service visits (i.e., unanticipated issues that require a service technician) are not included in the modeling assuming that the building personnel follow the maintenance and care correctly.

A lifetime of 35 years has been assumed, based on the statement in a boiler brochure: "It's not uncommon for Weil-McLain cast iron boilers to last 35 years or more."[18] Shah (2008) backs this number up as well.

At the end of life, it is assumed that the packaged unit is removed from the building and its metal parts are recycled, especially since the vast majority is recoverable metal, fully recyclable, and the equipment is easy to recover and remove from the building. See Footnote 14 for the recycling methodology used. The fiberglass is modeled as landfilled, and a distance of 48 km (30 mi) to the landfill in a heavy-duty diesel truck has been modeled. The landfill is modeled based on ecoinvent end of life waste management process data.

Fin Tube Radiation

Fin tube radiation is characterized by hot water heated by boiler and piped to "fin-tube" baseboard units mounted along walls. Fin tubes' heating elements are copper tubes with aluminum fins. Fin tube radiation is modeled for 5 heating applications which are used in the tool:

Table 4-34 Fin Tube Radiation Characteristics in BIRDS

Application of Fin Tube Radiation	Space being Heated: Area & Volume	Notes
Apartment building, forced hot water	1 858 m^2, 5 663 m^3 (20 000 ft^2, 200 000 ft^3)	
Apartment building, forced hot water	2 787 m^2, 8 495 m^3 (30 000 ft^2,300 000 ft^3)	
Commercial building, forced hot water	929 m^2, 2 832 m^3 (10 000 ft^2, 100 000 ft^3)	2 floors
Commercial building, forced hot water	9 290 m^2, 28 317 m^3 (100 000 ft^2, 1 000 000 ft^3)	3 floors
Commercial building, forced hot water	92 903 m^2, 283 168 m^3 (1 000 000 ft^2, 10 000 000 ft^3)	5 floors

---

[18] Boiler lifetime from Weil-McLain LGB brochure - C-805 (1111).

A material quantity surveyor estimated the bill of materials of the radiation system, based on the square footage and number of floors of the buildings using fin tube radiation (Dewhurst, 2012). For the apartment building applications, the estimates were based on a baseboard product by Slant/Fin (2004). For the commercial buildings, the Trane Architectural Hydronic Wall Fin (2001) was used. These were used because they were considered by the surveyor to be representative of current products used in today's buildings. Also, enough data were available from company literature to provide good estimates of the materials.

Upstream Materials Production through Manufacturing

The fin tube radiation bill of materials includes three aspects: fin tubes, their enclosures, and the distribution system which delivers steam from the boiler, through the building, to the rooms where the radiators provide heat. The boiler is described in another section. The figure below shows a close-up of the radiator (excludes the distribution system).

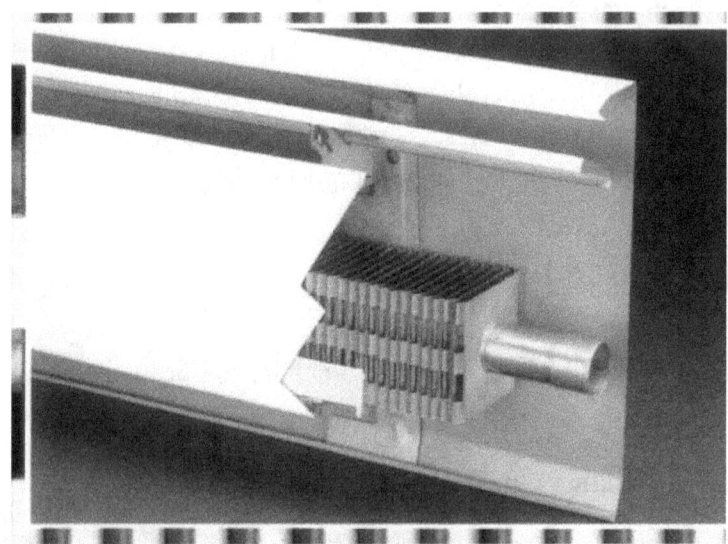

Figure 4-9 Fin Tube Radiator (Internal and External Components)[19]

The materials surveyor first defined the buildings' heating systems, including the boiler(s) and its output(s) and other elements that would be typical in that size of building. Then for each building type, he calculated the total lengths of fin tubes, horizontal piping, and risers needed, based on the intended use of the building (apartments or open commercial spaces) and size of the buildings. To do this, the surveyor took the total square meters (square feet) of building and divided by number of floors to get per floor square meters (square footage). Then he took the square root of per floor square meters (square footage) to determine perimeter wall lengths, and the horizontal distribution on each floor was assumed to run along this perimeter wall length on

---

[19] Slant/Fin, 2004.

each floor. The fin tube length to heat the buildings was estimated using 2.95 MJ/m (900 Btu/ft) of fin tube, based on the calculated heat demand for the building.

The distribution system included pipework for the risers, which run vertically from the boiler to each heated floor, the horizontal ring main on each floor, and the pieces connecting the distribution pipework to the fin tube radiators. The risers included the pipes, insulation, clamps and fittings. The horizontal ring main included the pipes, insulation, hangers, and fittings. Floor heights were assumed to be on average 3.5 m (11.5 ft), which was used to determine the length of vertical distribution. Fittings, connections, and installation components were calculated based on lengths. Gauge and weight charts for the sheet metals and other standard pipe tables were used (Greenheck, 2004). Table 4-36 summarizes the bill of materials for the three aspects.

Data for copper come from ICA (2012). Aluminum is modeled as a 50/50 mix of primary and secondary extruded aluminum which come from the U.S. LCI Database. Data for cold rolled steel and steel pipes come from World Steel Association (2011). Ecoinvent data were used for steel enameling and cast iron. The fiberglass insulation was modeled as described in the insulation section. Raw materials are modeled as transported to the manufacturing plant via diesel truck an assumed average distance of 805 km (500 mi).

Data for parts forming, sheet rolling, enameling, and pipe forming (drawing of pipes, steel, from ecoinvent) are included in raw materials production. No data were available to model assembly of the fin tubes.

Table 4-35 Fin Tube Radiation Summary of Materials

Assembly	Part/Material	Area 929 m² (10 000 ft²) Commercial 2 floors kg (lb)	Area 1 858 m² (20 000 ft²) Apartment 3 floors kg (lb)	Area 2 787 m² (30 000 ft²) Apartment 4 floors kg (lb)	Area 9 292 m² (100 000 ft²) Commercial 3 floors kg (lb)	Area 92 903 m² (1 000 000 ft²) Commercial 5 floors kg (lb)
Fin Tube	Pipe (copper)	66 (146)	102 (224)	142 (313)	439 (967)	3 192 (7 038)
	Fins (Al)	144 (318)	96 (212)	135 (297)	954 (2 104)	6 944 (15 310)
	Wallfin Installation Sheet (cold rolled steel)	2.0 (4.4)	4.4 (10)	6.2 (14)	13 (29)	97 (213)
	Wallfin Installation Bolts (carbon steel)	1.9 (4.3)	4.3 (9.4)	6.0 (13)	13 (28)	93 (205)
Enclosure	Sheet (cold rolled steel)	1 112 (2 452)	762 (1 680)	1 067 (2 353)	4 278 (9 432)	29 948 (66 024)
	Enclosure Installation Sheet (cold rolled steel)	174 (385)	329 (727)	461 (1 017)	672 (1 481)	4 704 (10 370)
	Paint (baked enamel)	8.8 (19)	6.0 (13)	8.4 (19)	34 (75)	2 367 (522)
	Enclosure Installation Bolts (carbon steel)	3.7 (8.1)	4.3 (9.4)	6.0 (13)	14 (31)	99 (219)
Distribution System	Pipe (sched 40 steel)	1 018 (2 244)	1 589 (3 505)	2 308 (5 089)	4 617 (10 179)	38 927 (85 820)
	Insulation (fiberglass)	75 (165)	118 (259)	169 (373)	296.(653)	2 206 (4 864)
	Riser Clamp (carbon steel)	4.5 (10)	7.5 (16)	10.5 (23)	11 (25)	69 (153)
	Hanger (carbon steel)	41 (90)	68 (150)	118 (259)	151 (333)	1 025 (2 262)
	Pipe fittings (black malleable Iron)	165 (163)	263 (580)	379 (835)	743 (1 639)	6 158 (13 575)
	Pipe (connect dist. to fin tube)	726 (1 600)	817 (1 800)	1 451 (3 200)	2 177 (4 800)	5 443 (12 000)
	Ready Rod for the hanger (carbon steel)	15 (34)	24 (53)	34 (76)	54 (119)	220 (484)

Transportation to the Building Site through End of Life

Transportation of the equipment to the building site is modeled to be an assumed average of 644 km (400 mi) by heavy-duty diesel fuel-powered truck.

Fin tube radiation lasts well over 40 years, so no replacement is needed for the 40-year study period. Maintenance of the radiators is minimal; bleeding the radiators may need to be done to let out air that has gotten trapped inside, and this is done manually so is not included in the model.

At the end of life, it is assumed that the fin tube radiators are removed from the building and recycled, especially since most of the parts are valuable recoverable metal, fully recyclable, and relatively easy to recover and remove from the building. The recycling methodology used is described in Footnote 14.

Additional Reference

Air-Conditioning, Heating, and Refrigeration Institute (AHRI), I=B=R Ratings Directory for Boilers, Baseboard Radiation, Finned Tube (Commercial) Radiation, Indirect-fired Water Heaters, Updated April 2009.

Packaged Units

Packaged units, or rooftop units, are air handlers designed to be installed for use on rooftops. An air handler is an appliance used to condition and circulate air through a building as part of the building's HVAC system. A split system is an air conditioner split by the condensing unit located on the outside of the building and the evaporator in a furnace or air handler inside the building. Four packaged units and one split system condensing unit were evaluated:

Table 4-36 Packaged Units Characteristics in BIRDS

Type of HVAC System	Application & Square Area	Designated Tonnage
Rooftop air conditioner – single zone	Department stores 929 m^2 (10 000 ft^2)	29.17
Rooftop air conditioner – multizone	Restaurants 279 m^2 (3 000 ft^2)	15.00
Rooftop air conditioner – multizone	Medical centers 2 323 m^2 (25 000 ft^2)	58.33
Rooftop air conditioner – multizone	Offices 2 323 m^2 (25 000 ft^2)	79.16
Air cooled condensing unit – split system	Schools and colleges 1 858 m^2 (20 000 ft^2)	76.66

Upstream Materials Production through Manufacturing

Rooftop Air Conditioner Production

Data for a commercial sized rooftop air conditioner were provided by a global producer of commercial HVAC equipment. The full bill of materials for the system was provided for BIRDS. The table below provides general data on the main components of the equipment but due to the proprietary nature of the data, the full details of all of the materials and the weight of the components have been removed.

Table 4-37 Packaged Units Main Components

Component name	Description & Material	Notes
Housing	Painted galvanized steel of varying thicknesses	
Condenser Coils	Aluminum	
Supply Fan	Mostly steel & a 10 HP motor	Assume 20 % motor, 80 % steel
Evaporator Coil	Copper tubing, Aluminum fin stock	
Condenser Fans	Includes 1-hp motors, fan blades, basket.	Assume 80 % motors, 20 % steel
Economizer Damper	Aluminum	
Control Panel	Sheet metal housing, electrical components	
Compressor	Steel, motors, copper	
Refrigerant	R410A	
Misc parts	Nuts, bolts, gaskets	Assume 90 % steel, 10 % rubber

This product requires a roof curb, which is a steel structure that supports HVAC equipment on top of a building. The curb is included in the model and is assumed to be galvanized steel.

Data for galvanized steel sheet for the housing come from World Steel Association (2011). The total weight for the compressor was given without a breakdown of the individual components. Data for a compressor was therefore modeled based on the percentages of the materials in a compressor reported in Biswas (2011):

Table 4-38 Compressor Bill of Materials

Material	Weight kg (lb)	%
Steel	15.5 (34)	9
Copper	7.5 (17)	4
Aluminum	3.0 (6.6)	2
Cast iron	141 (311)	84
Total	**167 (368)**	**100**

Aluminum for the condenser coils, evaporator coil, and economizer damper is modeled as a 50/50 mix of primary and secondary extruded aluminum which come from the U.S. LCI Database.

For the motors, a data set for an electric car motor, from ecoinvent, has been used as a guideline for the general material make-up for an electric motor. This data set includes the general

94

materials in the motor, including rolled steel (75 % of total), aluminum (approximately 16 %), and copper wire (9 %). U.S. LCI data sets were used for the aluminum, World Steel Association (2011) for the steel, and ICA (2012) for the copper.

Steel in the supply fan, compressor, and miscellaneous parts comes from cold rolled steel from World Steel Association (2011) and data for copper come from ICA (2012).

Control panel: The World Steel Association (2011) data for cold rolled steel was used for the sheet metal housing component of the control panel, which made up 80 %. Not enough data were available on the contents of this remaining part to model the electrical components. Synthetic rubber data come from ecoinvent.

R-410a data are based on a 50/50 share of difluoromethane (R-32) and pentafluoroethane (R-125). Due to lack of available production data on both of these chemicals, proxies were used. Trifluoromethane (HFC-23) was used as a proxy for difluoromethane and 1,1,difluoroethane (HFC 152a) was used as proxy for R125. While proxies were used for the production aspect of the chemicals, any release of these was based on the release of R-32 and R-125, not the proxy chemicals, so that ozone depletion impact remains zero and global warming potential impact is calculated appropriate to R-410A.

Raw materials are modeled as transported to the manufacturing plant via diesel truck an assumed average distance of 805 km (500 mi).

Production of Other Packaged Units

For lack of available data for the other products in this category, this bill of materials was normalized for each size of packaged unit included in BIRDS. The exception to this was the quantity of refrigerant used for each and the unit weights, both of which could be obtained or extrapolated for different models of rooftop air conditioners. The table below summarizes the data sources and details of the rooftop air conditioner units to meet the NIST specs:

Table 4-39 Packaged Unit Models

HVAC System	Product this is based on[20]	Weight kg (lb)	R-410a charge kg (lb)	Curb weight kg (lb)
Rooftop air conditioner – single zone (929 m² (10 000 ft²), 29.17 ton)	Maverick II commercial packaged rooftop system, 30 ton[21]	1 637 (3 610)	11.3 (25)	192 (423)
Rooftop air conditioner – multizone (279 m² (3 000 ft²), 15.00 ton)	Voyager Packaged Rooftop Air Conditioners, 15 ton[22]	1 035 (2 281)	5.2 (11.4)	192 (423)
Rooftop air conditioner – multizone (2 323 m² (25 000 ft²), 58.33 ton)	IntelliPak Packaged Rooftop Air Conditioners, 60 ton[23]	3 782 (8 338)	20.1 (44.4)	234 (515)
Rooftop air conditioner – multizone (2 323 m² (25 000 ft²), 79.16 ton)	IntelliPak Packaged Rooftop Air Conditioners, 75 ton[24]	4 011 (8 843)	25.2 (55.6)	277 (610)

Production of Air Cooled Condensing Unit – Split System

Bill of materials for the split system air cooled condensing unit were based on data for a Carrier Corporation air conditioner from Shah (2008). The mass of the Carrier Corporation 24ACR3 Comfort 13 Series (1-1/2 to 5 nominal tons) condenser unit in the article was normalized up to the mass of a commercial split system air-cooled condensing unit available on the market. The Trane split system condensing unit with remote evaporator chiller was chosen based on available data on weight and refrigerant quantity provided in company literature.

Legutko (2000, p.3) describes an evaporator as "a direct-expansion, finned, tubular coil that has refrigerant inside the tubes. A fan draws air across the finned exterior of the tubes and delivers it to the spaces being conditioned. Standard coil construction consists of copper tubes with aluminum fins mounted in a galvanized steel frame." The remote evaporator's bill of materials was based on Legutko (2000); the dimensions of the Trane remote evaporator and a gauge chart were used to estimate the galvanized steel frame, and assumptions were made on the remaining

---

[20] The packaged units in the table representing the NIST systems were chosen based on available data on weight and/or refrigerant quantity provided in company literature. The products listed do not imply endorsement or product preference or quality.

[21] Daikin McQuay Catalog 250-6: Maverick II commercial packaged rooftop systems: Heating & Cooling Models MPS015F – 075E, base weight from Table 36, curb weight from Table 37. Refrigerant data extrapolated based on other sources.

[22] Trane Product Catalog: Packaged Rooftop Air Conditioners: Voyager™ Cooling and Gas/Electric - 12½–25 Tons, 60 Hz, RT-PRC028-EN, Nov. 2011. Weight Table 88, refrigerant Table 1.

[23] Trane Product Catalog, Packaged Rooftop Air Conditioners IntelliPak™ — S*HL, S*HK - 20 - 130Tons —Air-Cooled Condensers — 60 Hz, RT-PRC036-EN, April 2012, weights from Table 84. Refrigerant is extrapolated based on other Trane systems.

[24] Trane Product Catalog, Packaged Rooftop Air Conditioners IntelliPak™ — S*HL, S*HK - 20 - 130Tons —Air-Cooled Condensers — 60 Hz, RT-PRC036-EN, April 2012, weights from Table 84. Refrigerant is extrapolated based on other Trane systems.

components.[25] The table below presents the bill of materials for the condenser unit and remote evaporator.

Table 4-40 Split System Bill of Materials

Unit	Material	Quantity in Residential System kg (lb)	Quantity in 80-ton Compressor kg (lb)
Condenser unit	Steel	78 (172)	1 189 (2 621)
	Galvanized steel	35 (77)	534 (1 176)
	Copper	17 (37)	259 (571)
	Aluminum	17 (37)	259 (571)
	Total	**147 (324)**	**2 241 (4 940)**
	Refrigerant R-22	6 (13)	n/a
	Refrigerant R-410a	n/a	26 (57)

Unit	Material	Quantity in Residential System kg (lb)	Quantity in Evaporator kg (lb)
Remote Evaporator	Galvanized steel	n/a	20 (45)
	Copper	n/a	54 (120)
	Aluminum	n/a	6 (13)
	Other steel parts	n/a	14 (30)
	Total	n/a	94 (208)

The material data for elements in the systems are described above.

Packaged Units Manufacturing

The commercial HVAC equipment manufacturer provided electricity for assembly (punching, shearing, forming parts), lights, and fans, natural gas usage, and propane for forklifts. Data were provided on a full-facility basis for 2010 and calculated on the basis of one unit, based on total production time for the various lines of HVAC equipment produced at the facility. For confidentiality purposes, these manufacturing data are not shared in this documentation, however, they have been included in the model, i.e., normalized for each size of unit in this category.

Transportation to the Building Site through End of Life

Transportation of the equipment to the building site is modeled to be an assumed average of 644 km (400 mi) by heavy-duty diesel fuel-powered truck.

---

[25] Trane Product Catalog: Split System Condensing Units, Fig. 41 (assumed 14 gauge), Tables 1, 11, 12.

It is assumed that a qualified service technician comes to the building site to check and service the unit every three years to ensure optimal performance and lifetime. It is assumed that the qualified technician is within a 24 km (15 mi) service radius. The 24 km (15 mi), driven in a gasoline-powered van, is shared amongst other service visits for that technician, assuming that the same technician is making more than one service call during that trip. Assuming the technician makes 5 service calls in one day, one-fifth of the impacts from driving 24 km (15 mi) are allocated to the product, or 4.8 km (3 mi). Data for a van come from ecoinvent. Unplanned service visits (i.e., unanticipated issues that require a service technician) are not included in the modeling assuming that the building personnel follow the maintenance and care correctly.

For the packaged units, a lifetime of 15 years has been assumed, based on an average of a 10 to 20 year life span for a non-coastal application.[26] For the split system unit, a lifetime of 20 years for the condenser unit and 25 years for the air handler have been assumed, based on the lifetimes reported by Shah (2008).

For this analysis, 0.5 % per year of the total refrigerant for each unit is assumed to leak.

End of Life

At the end of life, it is assumed that the packaged unit is removed from the building and recycled, especially since most of the parts are valuable recoverable metal, fully recyclable, and relatively easy to recover and remove from the building. The recycling methodology used is described in Footnote 14. The curb is not removed, as a new unit is installed in its place. The non-metal components are modeled as landfilled. A distance of 48 km (30 mi) to the landfill in a heavy-duty diesel truck has been modeled. The landfill is based on ecoinvent waste management process data.

Packaged Chillers
Air-cooled packaged chillers with scroll compressors and water-cooled packaged chillers with screw compressors were modeled for BIRDS. Additionally, these chillers were modeled with a fan coil unit, and the water cooled chillers included a cooling tower. These were evaluated for the following applications:

---

[26] Based on discussions with a manufacturer of packaged units.

Table 4-41 Packaged Chillers Characteristics in BIRDS

Type of HVAC System	Application & Square Area	Designated Tonnage
Air-cooled packaged chiller with scroll compressor	Medical centers or similar application, 1 858 m^2 (20 000 ft^2)	46.66
Air-cooled packaged chiller with scroll compressor	Medical centers or similar application, 3 716 m^2 (40 000 ft^2)	93.33
Water-cooled packaged chiller with screw compressor	Medical centers or similar application, 5 574 m^2 (60 000 ft^2)	140.00
Water-cooled packaged chiller with screw compressor	Offices, 5 574 m^2 (60 000 ft^2)	190.00
Water-cooled packaged chiller with screw compressor	Schools & colleges, 5 574 m^2 (60 000 ft^2)	230.00

Upstream Materials Production – Air-Cooled Chillers

No data were available for specifically the air-cooled packaged chiller, so the bill of materials used for the rooftop air conditioner in the previous section was used as a proxy (see Table 4-37). The data were customized to these packaged chillers in terms of the unit weights and quantity of refrigerant used, both of which could be obtained for models of air cooled packaged chillers that would meet the NIST specifications. The table below summarizes the data sources and details of the air-cooled packaged chillers to meet the NIST specs:

Table 4-42 Air Scroll Packaged Chiller Model

HVAC System	Product this is based on	Weight kg (lb)[*]	R-410a charge kg (lb)[*]
Air scroll packaged chiller with fan coil unit, 1 858 m^2 (20 000 ft^2), 46.66 ton	McQuay AGZ 045D (45 ton)	1551 (3420)	40 (88)
Air scroll packaged chiller with fan coil unit, 3 716 m^2 (40 000 ft^2), 93.33 ton	McQuay AGZ 090 (90 ton)	2539 (5605)	78 (172)

[*] McQuay International Product Catalog 611-1, Tables 33, 35

The data and assumptions used for the air scroll chillers, including production of raw materials and transportation of materials to the manufacturing facility, are found in the section above on rooftop air conditioners.

Upstream Materials Production – Water-Cooled Chillers

No data on water-cooled screw compressor chillers with a range of tonnage of 140 to 230 were available, so the bill of materials for a 2500-ton water-cooled centrifugal chiller from Institute for Construction and Environment (2011) was used as a proxy. It is acknowledged that water screw compressors and centrifugal chillers are different technologies but these data were used for lack of other data available and because they perform the function of cooling.

Data in the EPD were collected in 2010; the EPD provided the major assemblies of the unit in terms of weights of each, shown in the table below. Even though major assemblies were identified, the EPD provided the material composition in terms of percentage of the total mass, not in terms of materials within each assembly (next table). Thus, it was impossible to switch out assemblies to be able to better customize the data for the water screw chillers.

Table 4-43 Centrifugal Chiller Major Assemblies

Major Assemblies	kg	(lb)	% of total
Compressor	13 226	(29 157)	36
Compressor Install	2 300	(5 070)	6
Condenser	11 460	(25 265)	31
Controls	14	(31)	0.003
Economizer	303	(668)	1
Evaporator Tubes	7 463	(16 453)	20
Motors	165	(363)	0.4
Oil Tank	232	(511)	1
Purge	585	(1 290)	2
Shell	581	(1 281)	2
Unit Assembly	417	(919)	1
TOTAL	**36 745**	**(81 008)**	**100**

Table 4-44 Centrifugal Chiller Materials

Material / subcomponent	kg	(lb)	% of total
Steel	16 536	36 454	45
Copper	10 656	23 492	29
cast iron	8 084	17 822	22
Aluminum	735	1 620	2
Motor	735	1 620	2
TOTAL	**36 745**	**81 008**	**100**

Note: brass was listed but as 0 % so is not included in the table.

The material quantities were normalized to the weights of different models of water screw packaged chillers, and were customized in terms of the quantity of refrigerant used for each. The table below summarizes the data sources and details of the water-cooled packaged chillers to meet the NIST specs:

Table 4-45 Water-cooled Packaged Chiller Models

HVAC System	Product this is based on[*]	Weight kg (lb)	R-134a Charge (Both Circuits)
Water- cooled packaged chiller, medical centers, 5 574 m² (60 000 ft²), 140.00 ton	Trane RTWD 60 Hz standard efficiency (140 Ton)	3 016 (6 650)	118 (260)
Water- cooled packaged chiller, offices, 5 574 m² (60 000 ft²), 190.00 ton	Trane RTWD 60 Hz high efficiency (200 Ton)	3 705 (8 168)	124 (273)
Water- cooled packaged chiller, schools & colleges, 5 574 m² (60 000 ft²), 230.00 ton	Trane RTWD 60 Hz high efficiency (220 Ton)	4 079 (8 993)	165 (364)

* Trane Catalog Series R Helical Rotary Liquid Chillers, Tables 1, 67.

Data on cold rolled steel come from World Steel Association (2011) and data on copper come from ICA (2012). Ecoinvent data were used for the cast iron. Aluminum is modeled as a mix of primary and secondary extruded aluminum which comes from the U.S. LCI Database.

For the motors, a data set for an electric car motor, from ecoinvent, has been used as a guideline for the general material make-up for an electric motor. This data set includes the general materials in the motor, including rolled steel (75 % of total), aluminum (approximately 16 %), and copper wire (9 %). U.S. LCI data sets were used for the aluminum, World Steel Association (2011) for the steel, and ICA (2012) for the copper.

Data for 134a come from ecoinvent. Raw materials are modeled as transported to the manufacturing plant via diesel truck an assumed average distance of 805 km (500 mi).

Manufacturing

One manufacturer of water-cooled packaged chillers provided electricity and other energy data on a full-facility basis for 2010. From those data, the energy data were calculated on the basis of one unit, based on total production time for the various lines of HVAC equipment produced at the facility. For confidentiality purposes, these manufacturing data are not shared in this documentation, however, they have been included in the model.

Transportation to the Building Site through End of Life

Transportation of the equipment to the building site is modeled based on an assumed average of 644 km (400 mi) by heavy-duty diesel fuel-powered truck.

It is assumed that a qualified service technician comes to the building site to check and service the unit one time per year to ensure optimal performance and lifetime. It is assumed that the qualified technician is within a 24 km (15 mi) service radius. The 24 km (15 mi), driven in a gasoline-powered van, is shared amongst other service visits for that technician, assuming that the same technician is making more than one service call during that trip. Assuming the technician makes 5 service calls in one day, one-fifth of the impacts from driving 24 km (15 mi) are allocated to the product, or 4.8 km (3 mi). Data for a van come from ecoinvent. Unplanned service visits (i.e., unanticipated issues that require a service technician) are not included in the modeling assuming that the building personnel follow the maintenance and care correctly.

Bakane (2009) estimates a lifetime of 10 years for the scroll chiller and 17.5 years for the screw compressor (based on an average of 15 yrs to 20 yrs). For this analysis, 0.5 percent per year of the total refrigerant in each is assumed to leak, and this is modeled as recharged by the service technician.

At the end of life, it is assumed that these chillers are removed from the building and recycled, especially since most of the parts are valuable recoverable metal, fully recyclable, and relatively easy to recover and remove from the building. The recycling methodology used is described in Footnote 14. The non-metal components are modeled as landfilled; landfill data are come from ecoinvent. A distance of 48 km (30 mi) to the landfill in a heavy-duty diesel truck has been modeled.

Fan Coil

A fan coil is a device that has a heating or cooling coil and a fan, and is used to provide heat or air conditioning to the space in which it is installed. The bill of materials for the fan coil comes from Shah (2008, Table 3), as follows:

Table 4-46 Fan Coil Bill of Materials

Material	Weight kg (lb)
Steel	48 (106)
Galvanized steel	26 (57)
Copper	2 (4)
Total	**76 (168)**

Data for both cold rolled steel and galvanized steel come World Steel Association (2011) and data for copper come from ICA (2012). The fan coil is modeled as having a lifetime of 25 years, consistent with Shah (2008). At the end of life, it is assumed that the fan coil is recycled; the recycling methodology used is described in Footnote 14.

Cooling Tower

One cooling tower has been modeled to be used in conjunction with the water-cooled chillers in BIRDS. According to a Baltimore Aircoil Company (BAC) sales representative, the BAC Series 1500 cooling tower is the most appropriate for a commercial application needing a nominal tonnage in the range of 190 to 200 which generally represents an average for the water-cooled chillers in BIRDS). The bill of materials was compiled for the BAC 15201 (a single cell unit) from BAC Series 1500 Cooling Tower Specification (2010), which provided the following major components and subcomponents/parts.

Table 4-47 Cooling Tower Components and Parts

Component	Subcomponents / Parts
Cooling Tower	Panels, Panel edges, Panel finish
Cold Water Basin	Panels, Drain/Clean out connection, Make-up valve, Float shell, Float fill, Float connection
Water Outlet	Water outlet connection, Lift out strainers, Anti-vortexing device
Water Distribution System	Water inlet, Integral strainer, Open gravity type basin, Gravity flow nozzle, Basin weir, Metering orifice, Lift-off distribution covers
Fans	Fan blades, Fan cylinder, Fan guard
Bearings	Bearings, Grease, Seals
Fan Drive	Powerband
Sheaves	Sheaves
Fan Motor	Motor, Enclosure, Finish on winding, shafts and bearings
Fill and Drift Eliminators	Fill & Integral drift eliminators
Air Inlet Louvers	Air inlet louvers
Basin Water Level Sensing and Control	Enclosure, controls, water level, standpipe Venting, Standpipe mounting hardware
Accessories	Basin heaters, vibration cutout switch, basin sweeper piping, air intake option, platform, ladder, louver face platform, internal platform
Equipment Controls	Enclosed controls
Mounting Support	Support Structure

Any available data on these items, including specified surface areas, volumes, and weights were recorded. Some data on materials were provided by the literature while materials for other subcomponents and parts were assumed. The masses of the parts and materials were calculated using material densities and the manufacturer's published data. The following table presents the compiled bill of materials totaling the published mass of the BAC 15201 (1 942 kg (4 280 lb)).[27]

---

[27] BAC Product and Application Handbook –Series 1500 Engineering Data on Cooling Tower (no date), p. B58.

Table 4-48 Cooling Tower Bill of Materials

Material / component description	Mass kg	(lb)
Galvanized steel sheet	1 060	(2 337)
cold rolled steel	146	(321)
Aluminum	38	(84)
Cast iron	2.3	(5.0)
Equipment control assemblies	6.4	(14)
electric heater	6.8	(15)
Epoxy polyester	3.7	(8.2)
HDPE	1.8	(4.0)
Heavy-gauge steel	161	(354)
Polypropylene	1.8	(4.0)
Polystyrene	0.1	(0.2)
PVC	379	(835)
Stainless Steel	8.7	(19)
Fiberglass reinforced polyester	109	(239)
Motor	18	(40)
Total	**1 942**	**(4 280)**

Data for the galvanized steel sheet, cold rolled steel, and heavy-gauge steel come from World Steel (2011). U.S. LCI Database provided data for HDPE, PVC, general purpose polystyrene, polypropylene, and aluminum (assumed 50/50 primary and secondary).

Ecoinvent provided data on the cast iron, control assemblies, stainless steel, epoxy resin (for the epoxy polyester), and fiberglass reinforced polyester. For the motor, a data set for an electric car motor, from ecoinvent, has been used as a guideline for the general material make-up of an electric motor. This data set includes the general materials in the motor, including rolled steel (75 % of total), aluminum (approximately 16 %), and copper wire (9 %).

At the end of life, it is assumed that the cooling tower is removed from the building and recycled, especially since most of the materials are valuable recoverable metal and relatively easy to recover and remove from the building. The recycling methodology used is described in Footnote 14.

4.5.5 Overhangs

Overhangs can be used to block the entry of direct sunshine into a building throughout the warm summer months and allow in direct sunlight during the winter. The optimal application of an overhang can be calculated based on latitude, to maximize shade during the warm summer months and maximize sunshine in the winter. For example, this can be accomplished using a 45.7 cm (18 in) overhang in Philadelphia where the sun is 85 % off the horizon in the summer at noon and drops to 25 % over the winter.[28] Specifications for an overhang on a particular building in one part of the country may vary greatly from that in another part.

For BIRDS, 0.09 m2 (1 ft^2) of an overhang has been modeled for the LEC scenario. Overhangs are placed on the east, west, and south sides of the building for each floor in Climate Zone 1 through Climate Zone 5 because these warmer climates are the zones that benefit from blocking solar radiation. The flow diagram below presents the general system boundaries for the overhang category, as it is modeled for BIRDS.

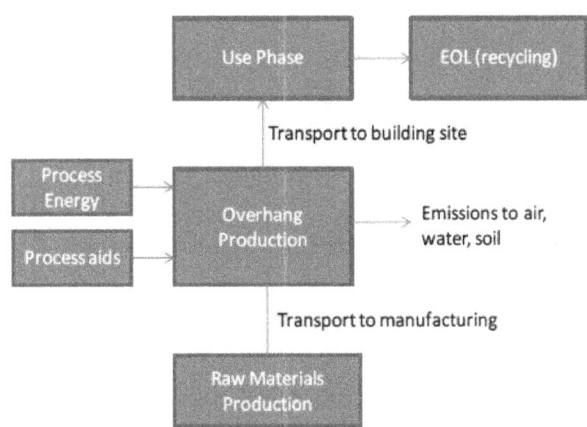

Figure 4-10 Overhang System Boundaries

Upstream Materials Production through Manufacturing

An aluminum overhang used on commercial buildings, including on curtain walls and storefront windows, has been modeled. The overhang modeled for BIRDS is meant to be an average, not one specifically designed for every building in every geographical location. The data are based on a product manufactured by Kawneer Company, Inc., the Outrigger System Versoleil SunShade, as shown in the figure below. According to a Kawneer architectural representative, the outrigger side piece, the fascia (end-piece), louvers (blades), and hardware that affix to the building are approximately 14.9 kg/m (10 lb/ft) for the 76.2 cm (30 in) deep outrigger model. For 1 ft^2, this amounts to 1.8 kg (4.0 lb).

Aluminum is modeled based on a 50/50 mix of primary and secondary extruded aluminum from the U.S. LCI Database. Manufacturing includes forming of the parts, and assembly. Parts forming data come from ecoinvent. Given its relatively simple design as shown in the figure below (taken from Kawneer, 2011), assembly is not expected to be a significant part of manufacturing impacts.

---

[28] "Southern Overhangs", retrieved July 2013 from
http://www.greenandsave.com/cooling/c/southern_overhangs.html.

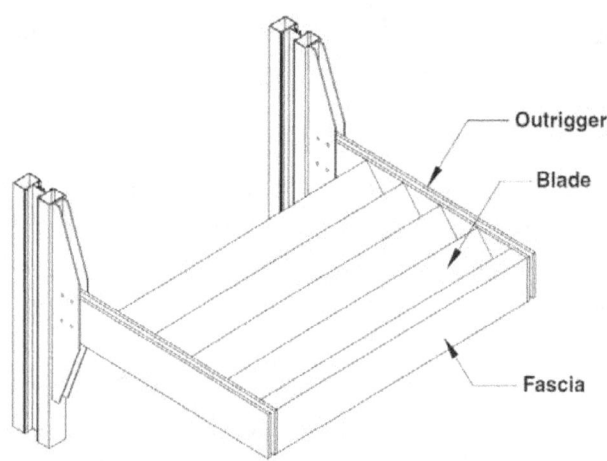

Figure 4-11 Outrigger System

Transportation to the Building Site through End of Life

Transportation of the equipment to the building site is modeled based on an assumed average of 805 km (500 mi) by heavy-duty diesel fuel-powered truck. Installation includes welding to affix the piece onto the building. The product lifetime is over 40 years so no replacement is necessary. Any maintenance needed, such as cleaning or refinishing the overhang, is not included.

At the end of life, it is assumed that the aluminum overhang is removed from the building and recycled, especially since aluminum is a valuable, fully recyclable metal. See Footnote 14 for the recycling methodology used.

4.5.6 Daylighting

A daylighting dimming control system is modeled for all of the climate zones for the LEC scenario. A dimming control system is used to adjust the level of electric light in a building when enough daylight can compensate for the electric source. Ander (2012) describes dimming controls as "continuously adjust[ing] electric lighting by modulating the power input to lamps to complement the illumination level provided by daylight." A photo sensor is used to sense the natural daylight and match that to the lighting system output.

The system modeled in BIRDS is a dimming control system that can control several lighting zones. The functional unit is one square foot of light adjusted by a dimming system, based on the use of approximately 10 fixtures per 92.9 m^2 (1 000 ft^2).

Daylighting systems come in many configurations and range from very complex systems to very simple devices. The WattStopper LightSaver daylighting sensors and control system was chosen for data collection since it is being put into use in a range of buildings and applications and is a fairly commonly used system.[29] The components included in the dimming system modeled for BIRDS include:[30]

- LightSaver LCD-203 Dimming Controller which "provides automatic dimming control for fluorescent and high-intensity discharge (HID) fixtures. It is an open loop controller providing up to three zones of control from a single photocell."
- LightSaver LS-290C Photosensor which "provides the daylight data necessary for operation of the LCD-203 daylighting control systems."
- LightSaver BT-203 Power Pack which "powers the LightSaver LCD-203 control module."
- LightSaver Wall Switches which "allow occupants to temporarily override WattStopper's automatic multizone daylighting control systems."

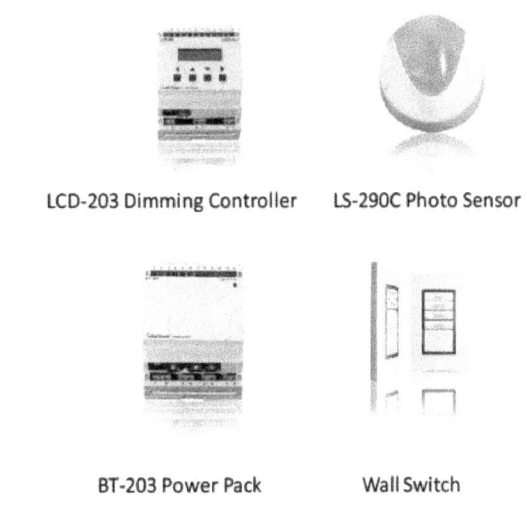

LCD-203 Dimming Controller   LS-290C Photo Sensor

BT-203 Power Pack   Wall Switch

Source of each picture: http://www.wattstopper.com

Figure 4-12 Components of the Dimming System

Three photo sensors and wall switches are used for one dimming controller and power pack. To meet the goal of approximately 10 fixtures per 93 m^2 (1000 ft^2), three sets of systems have been modeled, i.e., three dimming controllers and power packs and nine photo sensors and wall switches. These were then normalized to 0.09 m^2 (1 ft^2) of space.

The system requires the use of dimmable ballasts. In a retrofit application or upgrade where daylighting was not originally specified, then dimming ballasts need to be ordered with the dimming system and sockets in the existing fixture (fluorescent fixtures) need to be changed to accommodate for the dimming. However, in a new building in which a dimming system is

---

[29] Based on conversation with a representative at Stoneway Electric, Seattle, WA, April 2012.
[30] Wattstopper product descriptions, accessed June 2013 from: http://www.wattstopper.com/categories/daylighting-sensors-and-controls/systems.aspx.

specified, dimming ballasts are already installed. This system does not include the production of dimming ballasts in the new buildings. The flow diagram below presents the general system boundaries for the daylighting system, as it is modeled for BIRDS.

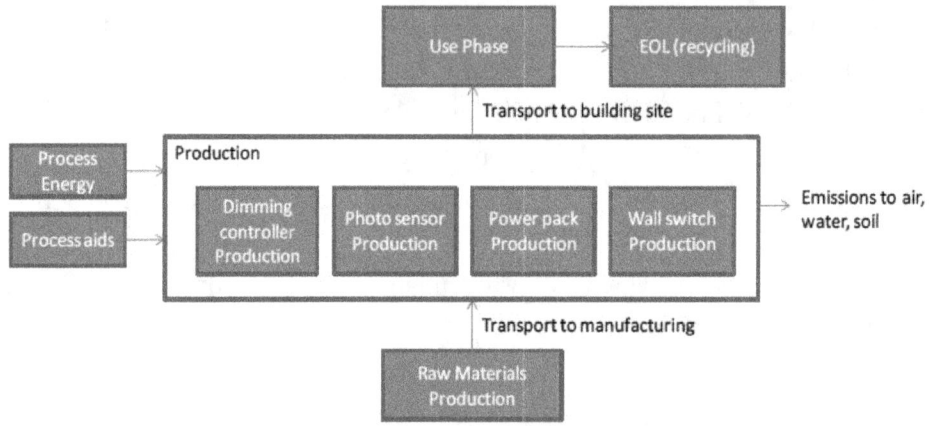

Figure 4-13 Daylighting System Boundaries

Upstream Materials Production through Manufacturing

The bill of materials for each component was obtained by way of complete tear-downs of each, presented in the tables below.

Table 4-49 Dimmer Controller LCD-203 Parts and Materials

Part Description	Dimension	MAss g	(lb)	Material
RJ12 cable	30.2 cm x 0.7 cm (77 in x 1.8 in)	8.63	0.019	Cable
Case		54.34	0.120	Polycarbonate
Controller buttons		1.27	0.003	Rubber
Circuit board	8.7 cm x 6.7 cm (22 in x 17 in)	57.59	0.127	Polychlorinated biphenyls
Circuit board	6.7 cm x 6.7 cm (17 in x 17 in)	43.58	0.096	Polychlorinated biphenyls
Circuit board	6.7 cm x 3.8 cm (17 in x 9.7 in)	11.18	0.025	Polychlorinated biphenyls
Display	4.2 cm x 2.0 cm (11 in x 5.1 in)	11.04	0.024	Liquid crystal display
Ribbon cable	7.5 cm x 1.7 cm (19 in x 4.3 in)	1.13	0.002	flat cable

Table 4-50 Photo Sensor LS-290C Parts and Materials

Part Description	Dimension	Mass g	(lb)	Material
Circuit board	4.6 cm x 3.4 cm (11.7 in x 8.6 in)	7.78	0.017	Polychlorinated biphenyls
White opaque cover		4.67	0.010	Polycarbonate
Translucent lense		8.21	0.018	Polycarbonate
Base		4.67	0.010	Polycarbonate
Adhesive circle		0.28	0.001	Acrylic

Table 4-51 Power Pack BT-203 Parts and Materials

Part Description	Dimension	Mass g	(lb)	Material
Case		54.19	0.119	Polycarbonate
Circuit Board 1	6.6 cm x 6.6 cm (16.8 in x 16.8 in)	33.25	0.073	Polychlorinated biphenyls
Circuit Board 2	8.7 cm x 6.7 cm (22 in x 17 in)	117.02	0.258	Polychlorinated biphenyls

Table 4-52 Wall Switch LS-5C Parts and Materials

Part Description	Dimension	Mass g	(lb)	Material
Cover plate		19.95	0.044	Polycarbonate
Push buttons		1.40	0.003	Polycarbonate
Mounting screws		3.11	0.007	Steel
Circuit board mounting screws		1.13	0.002	Steel
Circuit board	7.0 cm x 4.1 cm (17.8 in x 10.4 in)	20.09	0.044	Polychlorinated biphenyls
Back plate cover		19.81	0.044	Polycarbonate

Materials

It should be noted that in the vast majority of cases, there were no material identifiers on the parts, so often a best-guess was made as to the actual material. Data for steel come from World Steel Association (2011). Data for acrylic binder, synthetic rubber, and polycarbonate come from ecoinvent, as does injection molding of the plastic parts. Ecoinvent electronics data modules were used for the electronics parts, and the following data sets were used for the specified parts:

Table 4-53 Ecoinvent Data Sets for Electronics Parts

Part	Corresponding Ecoinvent Data Set(s)	Unit
Circuit board (PCB)	Printed wiring board, surface mount	Square area
Cable	Cable, network cable, category 5, without plugs	Length
	Plugs, inlet and outlet, for network cable	Per piece
LCD display	LCD glass	Mass
Ribbon cable	Cable, ribbon cable, 20-pin, with plugs	Mass

Raw materials are modeled as transported to the manufacturing plant via combination diesel truck an assumed average distance of 805 km (500 mi).

Manufacturing data are included in the electronics components data sets, and parts forming (i.e., injection molding of plastics) is included with much of the raw materials production data. Assembly of these pieces of equipment is not included, however.

Transportation to the Building Site through End of Life

It is assumed that these components are manufactured in Asia, and Shanghai, China has been assumed. The transportation distance of 10 600 km (5 720 mi) by ocean freighter from Shanghai to Los Angeles, and then an average of 2 414 km (1 500 mi) by heavy-duty (combination) diesel fuel-powered truck to users in the U.S. has been modeled.

Lighting loads in the buildings are discussed in Chapter 3. The energy simulations for the prototype buildings take into account lighting density, in watts per unit of conditioned floor area, appropriate to the building design (e.g., edition of *ASHRAE 90.1* or LEC). The lighting load schedules take into account occupancy usage appropriate to the building types. For LEC cases, the dimming system is incorporated in these calculations.

Lighting systems, including daylighting controls for the LEC design, are assumed to be replaced every 20 years. At end of life, the components are removed from the building and are recycled with other electronic waste.

5 Economic Performance Measurement

When a decision maker wants a dollar measure of cost effectiveness and cash flows are primarily costs, the most appropriate method for measuring the economic performance of a building is the life-cycle cost (LCC) method (ASTM, 2012). BIRDS follows the ASTM International standard method for life-cycle costing of building-related investments (ASTM, 2010). It involves calculating a cost's present value (PV) by discounting its future value into 2009 dollars based on the year the cost occurs and the assumed discount rate. The formulas and discount factors used to calculate the present values will vary depending on the type of cost. The different cost types and related formulas, discount factors, and data sources are described below. [31]

5.1 First Cost

5.1.1 Approach

The first costs of a building are the total costs of constructing a building in a particular city. First costs include costs for labor, materials, equipment, overhead, and profit. The construction costs for a prototype building are estimated by summing the costs of the baseline building (C_{NatAvg}) and the changes in costs required to meet the prototype building design (ΔC_x), adjusted for location-related cost variation as well as contractor and architectural profits. Both the baseline building costs and component cost estimates are based on national average construction cost data, and must be adjusted with the 2009 RS Means *CostWorks City Indexes* to control for local material and labor price variations in the 228 locations for which energy simulations are run. The "weighted average" city construction cost index (I_{WAvg}) is used to adjust the costs for the baseline prototypical building while "component" city indexes (I_x) are used to adjust the costs for the change in component designs. The formula below shows the indexed construction cost (C_{Index}) calculation.

$$C_{Index} = \left(C_{NatAvg} * I_{WAvg}\right) + \left(\Delta C_{HVAC} * I_H\right) + \left(\Delta C_{Wall} + \Delta C_{Roof}\right) * I_T + \left(\Delta C_{Light} + \Delta C_{Daylight}\right) * I_E$$
$$+ \left(\Delta C_{Window} + \Delta C_{Overhang}\right) * I_O$$

Where C_{Index} = Indexed construction costs
        C_{NatAvg} = National average construction costs
        ΔC_{HVAC} = Change in HVAC costs
        ΔC_{Wall} = Change in wall insulation costs
        ΔC_{Roof} = Change in roof insulation costs
        ΔC_{Light} = Change in lighting costs
        $\Delta C_{Daylight}$ = Change in daylighting system costs
        ΔC_{Window} = Change in window costs
        $\Delta C_{Overhang}$ = Change in overhang costs
        I_H = "Fire Suppression, Plumbing, & HVAC" cost index
        I_T = "Thermal and Moisture Protection" cost index
        I_E = "Electrical, Communications, & Utilities" cost index
        I_O = "Openings" cost index

---

[31] See Kneifel (2012) for additional details on the cost data used in the BIRDS database.

Once the indexed construction costs of the building are calculated, it is necessary to adjust for the contractor and architect profits by multiplying the costs by the contractor "mark-up" rate (I_M), assumed to be 25 %, and then the architectural fees rate (I_A), assumed to be 7 %, as shown in the following equation.

$$C_{First} = \left(C_{Indexed} * (1 + I_M)\right) * (1 + I_A)$$

These mark-up rates are based on the default values used by the RS Means *Square Foot Cost Estimator (SFCE)*. The marked-up, indexed construction costs are the first costs of constructing the prototype building in the particular city (C_{First}).

5.1.2 Data

Building construction costs are obtained from the RS Means *CostWorks* online databases. The costs of a prototypical building are estimated by the RS Means *CostWorks SFCE* to obtain the default costs for each BIRDS building type for each component. The RS Means default building is the baseline used to create a building that is compliant with each of the five energy efficiency design alternatives: *ASHRAE 90.1-1999, ASHRAE 90.1-2001, ASHRAE 90.1-2004, ASHRAE 90.1-2007*, and the higher efficiency "Low Energy Case" (LEC) design based on *ASHRAE 189.1-2009*. The RS Means default buildings are adapted to match the five prototype building designs by using the RS Means *CostWorks Cost Books* databases.

Five components -- roof insulation, wall insulation, windows, lighting, and HVAC efficiency -- are changed to make the prototypical designs compliant with *ASHRAE 90.1-1999, -2001, -2004, and -2007*. A summary of the minimum requirement ranges, excluding HVAC efficiency, for each building design are shown in Table 5-1. The windows are selected to meet the minimum window characteristics (U-factor, solar heat gain coefficient (SHGC), and visible transmittance (VT)) required by the building design at the lowest possible cost. The lighting density in watts per unit of conditioned floor area is adjusted to meet each standard edition's requirements.

Table 5-1 Energy Efficiency Component Requirements for Alternative Building Designs

Design Component	Parameter	Units	ASHRAE 90.1-1999	ASHRAE 90.1-2001	ASHRAE 90.1-2004	ASHRAE 90.1-2007	Low Energy Case*
Roof Insulation	R-Value	$m^2 \cdot K/W$ ($ft^2 \cdot F \cdot h/Btu$)	1.7 to 4.4 (10.0 to 25.0)	1.7 to 4.4 (10.0 to 25.0)	2.6 to 3 5 (15.0 to 20.0)	2.6 to 3.5 (15.0 to 20.0)	4.4 to 6.2 (25.0 to 35.0)
Wall Insulation	R-Value	$m^2 \cdot K/W$ ($ft^2 \cdot F \cdot h/Btu$)	0.0 to 3.8 (0.0 to 21.6)	0.0 to 3.8 (0.0 to 21.6)	0.0 to 2.7 (0.0 to 15.2)	0.0 to 2.7 (0.0 to 15.2)	0.7 to 5.5 (3.8 to 31.3)
Windows	U-Factor	$W/(m^2 \cdot K)$ ($Btu/(h \cdot ft^2 \cdot F)$)	1.42 to 7.21 (0.25 to 1.27)	1.42 to 7.21 (0.25 to 1.27)	1.99 to 6.47 (0.35 to 1 14)	2.50 to 6.47 (0.44 to 1.14)	1 97 to 6.42 (0.35 to 1.13)
	SHGC	Fraction	0.14 to NR†	0.14 to NR†	0.17 to NR†	0.25 to NR	0 25 to 0.47
Lighting	Power Density	W/m^2 (W/ft^2)	14.0 to 20.5 (1.3 to 1.9)	14.0 to 20.5 (1.3 to 1.9)	10.8 to 16.1 (1.0 to 1 5)	10.8 to 16.1 (1.0 to 1.5)	8.6 to 16.1 (0.8 to 1.5)
Overhangs			None	None	None	None	Zones 1 to 5
Daylighting			None	None	None	None	Zones 1 to 8

†North facing SHGC requirements are less restrictive than the requirements for the other 3 orientations.

* Low Energy Case design requirements are taken from the EnergyPlus simulations.

NR = No Requirement for one or more climate zones. By definition, the value of SHGC cannot exceed 1.0.

The LEC design increases the thermal efficiency of insulation and windows beyond *ASHRAE 90.1-2007*, further reduces the lighting power density, and adds daylighting and window overhangs. The lighting density of the lighting system is decreased by first increasing the efficiency of the lighting system and then decreasing the number of fixtures in the lighting system.[32] Daylighting is included for all building types and climate zones. Overhangs are placed on the east, west, and south sides of the building for each floor in Climate Zone 1 through Climate Zone 5 because these warmer climates are the zones that benefit from blocking solar radiation.[33]

Since the design of the BIRDS database, *ASHRAE 90.1-2010* has been finalized and published. Table 5-2 gives a perspective of how the *ASHRAE 90.1-2007* and LEC designs compare to *ASHRAE 90.1-2010*. In general, the requirements for *ASHRAE 90.1-2010* are more strict than *ASHRAE 90.1-2007* and less strict than the LEC design. The key exception is the lighting power density requirements for mid-rise and high-rise residential buildings, and wall insulation requirements for steel-framed mid-rise and high-rise residential buildings, where the requirements for *ASHRAE 90.1-2010* are more restrictive than for the LEC design.

---

[32] First, incandescent lighting is replaced with compact fluorescent lighting while typical T-12 fluorescent tube lighting is replaced with more efficient T-8 fluorescent tube lighting to decrease the lighting density of the lighting system. Second, the number of fixtures is reduced to meet the remainder of the required reduction in watts per unit of floor area. Increasing the efficiency of the lighting increases the costs of construction. The first approach increases first costs while the second approach decreases first costs for the lighting system. This approach is based on Belzer et al. (2005) and Halverson et al. (2006).

[33] Overhang cost source is Winiarski et al. (2003)

Table 5-2 Energy Efficiency Component Requirements for the ASHRAE 90.1-2007, ASHRAE 90.1-2010, and LEC Designs

Design Component	Parameter	Units	ASHRAE 90.1-2007	ASHRAE 90.1-2010	Low Energy Case*
Roof Insulation	R-Value	$m^2 \cdot K/W$ ($ft^2 \cdot$ F·h/Btu)	2.6 to 3.5 (15.0 to 20.0)	2.6 to 3.5 (15.0 to 20.0)	4.4 to 6.2 (25.0 to 35.0)
Wall Insulation	R-Value	$m^2 *K/W$ ($ft^2 \cdot$ F·h/Btu)	0.0 to 2.7 (0.0 to 15 2)	0.0 to 2.7 (0.0 to 25.0)	0.7 to 5.5 (3.8 to 31.3)
Windows	U-Factor	$W/(m2 \cdot K)$ ($Btu/(h \cdot ft^2 \cdot$ F))	2.50 to 6.47 (0.44 to 1.14)	1.99 to 6.47 (0.35 to 1.2)	1.97 to 6.42 (0.35 to 1.13)
	SHGC	Fraction	0 25 to NR	0.25 to NR†	0.25 to 0.47
Lighting	Power Density	W/m^2 (W/ft^2)	10.8 to 16.1 (1.0 to 1.5)	10.8 to 16.1 (0.6 to 1.4)	8.6 to 16.1 (0.8 to 1.5)
Overhangs			None	None	Zones 1 to 5
Daylighting			None	Zones 1 to 8	Zones 1 to 8

* Low Energy Case design requirements are taken from the EnergyPlus simulations, and are based on *ASHRAE 189.1-2009*.
NR = No Requirement for one or more climate zones. By definition, the value of SHGC cannot exceed 1.0.

Table 5-3 summarizes the HVAC efficiency requirements for each building design option across the different types of HVAC equipment.[34] This first version of BIRDS assumes that cooling equipment is run on electricity while heating equipment is run on natural gas. Note that the LEC design assumes the same efficiency as *ASHRAE 90.1-2007*. The most significant increases in HVAC efficiency requirements occur between *ASHRAE 90.1-1999* and *ASHRAE 90.1-2001* except for rooftop packaged units, which have consistently increasing requirements across multiple *ASHRAE 90.1 Standard* editions.

Table 5-3 HVAC Energy Efficiency Requirements for Alternative Building Designs

HVAC Type	Equipment Type	Unit	ASHRAE 90.1-1999	ASHRAE 90.1-2001	ASHRAE 90.1-2004	ASHRAE 90.1-2007	Low Energy Case
Cooling	Rooftop Packaged Unit	EER	8.2 to 9.0	9.0 to 9.9	9.2 to 10.1	9.5 to 13.0	9.5 to 13.0
	Air-Cooled Chiller	COP	2.5 to 2.7	2.8	2.8	2.8	2.8
	Water-Cooled Chiller	COP	3.80 to 5.20	4.45 to 5.50	4.45 to 5.50	4.45 to 5.50	4.45 to 5.50
	Split System with Condensing Unit	EER	8.7 to 9.9	9.9 to 10.1	10.1	10.1	10.1
Heating	Hot Water Boiler	E_t	75 % to 80 %	75 % to 80 %	75 % to 80 %	75 % to 80 %	75 % to 80 %
	Furnace	E_t	80 %	75 % to 80 %	75 % to 80 %	75 % to 80 %	75 % to 80 %

Assume that $E_c = 75\% E_t$ and $AFUE = E_t$, where E_c = combustion efficiency; E_t = thermal efficiency; AFUE = Annual Fuel Utilization Efficiency
EER = Energy Efficiency Ratio
COP = Coefficient of Performance
Note: Efficiency requirement ranges are based on the system sizes calculated in the whole building energy simulations.

The HVAC system size varies across the five building designs because changing the thermal characteristics of the building envelope alters the heating and cooling loads of the building. The

---

[34] This study does not account for new HVAC efficiency requirements set by federal regulations.

EnergyPlus whole building energy simulations "autosize" the HVAC system to determine the appropriate system size to efficiently maintain the thermal comfort and ventilation requirements. For each building design, the HVAC cost for the default HVAC system is replaced with the cost of the "autosized" HVAC system. An HVAC efficiency cost multiplier is used to adjust the HVAC costs in accordance with the standard efficiency requirements shown in Table 5-3.

5.2 Future Costs

5.2.1 Approach

Building maintenance, repair, and replacement (MRR) costs are discounted to equivalent present values using the Single Present Value (SPV) factors for future non-fuel costs reported in Rushing and Lippiatt (2009). These factors are calculated using the DOE Federal Energy Management Program (FEMP) 2009 real discount rate for federal energy conservation projects (3 %). Table 5-4 reports the SPV factors used in BIRDS. The MRR costs for each year ($C_{MRR,i}$) are multiplied by the SPV for that year and then summed and indexed to determine the total present value MRR costs (C_{MRR}).

Table 5-4 2009 SPV Discount Factors for Future Non-Fuel Costs, 3 % Real Discount Rate

Yrs	SPV Factor	Yrs	SPV Factor	Yrs	SPV Factor	Yrs	SPV Factor
1	0.971	11	0.722	21	0.538	31	0.400
2	0.943	12	0.701	22	0.522	32	0.388
3	0.915	13	0.681	23	0.507	33	0.377
4	0.888	14	0.661	24	0.492	34	0.366
5	0.863	15	0.642	25	0.478	35	0.355
6	0.837	16	0.623	26	0.464	36	0.345
7	0.813	17	0.605	27	0.450	37	0.335
8	0.789	18	0.587	28	0.437	38	0.325
9	0.766	19	0.570	29	0.424	39	0.316
10	0.744	20	0.554	30	0.412	40	0.307

The electricity and natural gas use predicted by the building's energy simulation is used as the annual energy use of the building for each year of the selected study period. Electricity and natural gas prices are assumed to change over time according to U.S. Energy Information Administration forecasts from 2009 to 2039. These forecasts are embodied in the FEMP Modified Uniform Present Value Discount Factors for energy price estimates (UPV*) reported in Rushing and Lippiatt (2009).[35] Multiplying the annual electricity costs and natural gas costs by the associated UPV* value for the study period of interest estimates the present value total electricity costs (C_{Elect}) and natural gas costs (C_{Gas}). The discount factors vary by Census region, end use, and fuel type.

Total present value future costs (C_{Future}) is the sum of present value location-indexed MRR costs and present value energy costs, as shown in the following equation:

---

[35] Since the U.S. Energy Information Administration forecasts end at year 30, the escalation rates for years 31-40 are assumed to be the same as for year 30.

$$C_{Future} = C_{MRR} + C_{Elect} + C_{Gas}$$

5.2.2 Data

Component and building lifetimes and component repair requirements are based on data from Whitestone (2008). Building service lifetimes are assumed constant across climate zones: apartment buildings last for 65 years; dormitories for 44 years; and hotels, schools, office buildings, retail stores, and restaurants for 41 years.

Building component maintenance, repair, and replacement (MRR) rates are from Kneifel (2010) and Kneifel (2011a). Insulation and windows are assumed to have a lifespan greater than 40 years and have no maintenance requirements. Insulation is assumed to have no repair costs. Windows have an assumed annual repair cost equal to replacing 1 % of all window panes, with costs that vary depending on the required window specifications. The heating and cooling units have different lifespans and repair rates based on climate, ranging from 4 to 33 years for repairs and 13 to 50 years for replacements.

Maintenance, repair, and replacement cost data are collected from two sources. The total maintenance and repair costs per square foot of conditioned floor area (minus the HVAC maintenance and repair costs) represent the baseline MRR costs per unit of floor area, which occur for a building type regardless of the energy efficiency measures incorporated into the design. These data are collected from Whitestone (2008), which reports average maintenance and repair costs per unit of floor area by building component for each year of service life for each building type. The building types in Whitestone do not match exactly to the 11 building types selected for this study, and the most comparable profile is selected.

RS Means *CostWorks* is the source of MRR costs for the individual components for which MRR costs change across alternative building designs, which in this analysis are the HVAC system, lighting system, and windows. Lighting systems, including daylighting controls for the LEC design, are assumed to be replaced every 20 years. The HVAC system size varies based on the thermal performance of the building design, which results in varying MRR costs because smaller systems are relatively cheaper to maintain, repair, and replace.

Future MRR costs are discounted to equivalent present values using the Single Present Value (SPV) factors for future non-fuel costs reported in Rushing and Lippiatt (2009), which are calculated using the U.S. Department of Energy's 2009 real discount rate for energy conservation projects (3 %).

Annual energy costs are estimated by multiplying annual electricity and natural gas use predicted by the building's energy simulation by the average state retail commercial electricity and natural gas prices, respectively. Average state commercial electricity and natural gas prices for 2009 are collected from the Energy Information Administration (EIA) *Electric Power Annual State Data Tables* and *Natural Gas Navigator*, respectively.

5.3 Residual Value

A building's residual value is its value remaining at the end of the study period. In life-cycle costing it is treated as a negative cost item. In BIRDS, it is estimated in three parts, for the building (excluding HVAC and lighting), the HVAC system, and the lighting system, based on the approach defined in Fuller and Petersen (1996). The building's residual value is calculated as the building's location-indexed first cost multiplied by one minus the ratio of the study period to the service life of the building, discounted from the end of the study period. For example, if a building has first costs (excluding HVAC and lighting) of $1 million, a 41 year service life, and the study period length is 10 years, the residual value of the building in year 10 (excluding HVAC and lighting) is $1\ 000\ 000 * \left(1 - \frac{10}{41}\right) = \$756\ 098$. The value is then discounted into present value terms.

Because they may be replaced during the study period, residual values for the HVAC and lighting systems are computed separately. Residual values for the HVAC system components, which have different lifespans across locations, are computed for each location. The remaining "life" of the HVAC equipment is determined by taking its service life minus the number of years since its last installation (as of the end of the study period), whether it occurred during building construction or replacement. The ratio of remaining life to service life is multiplied by the location-indexed installed cost of the system and discounted from the end of the study period. For example, assume an HVAC system's installed costs are $100 000 with a service life in the selected location of 8 years, and a 10-year study period length. After one replacement, the system is 2 years old at the end of the study period, leaving 6 years remaining in its service life. The residual value in year 10 is $100\ 000 * \frac{6}{8} = \$75\ 000$.

While its lifespan does not vary across locations, the residual value for the lighting system is computed in a similar manner. The lighting system service life is 20 years.

The total residual value of the building and its HVAC and lighting systems, multiplied by the SPV factor for the number of years in the study period, estimates the present value residual value ($C_{Residual}$).

5.4 Life-Cycle Cost Analysis

The total life-cycle cost of a prototype building (C_{LCC}) is the sum of the present values of first cost and future costs minus the residual value as shown in the following equation:

$$C_{LCC} = C_{First} + C_{Future} - C_{Residual}$$

LCC analysis of buildings typically compares the LCC for a "base case" building design to the costs for alternative, more energy efficient building design(s) to determine if future operational savings justify higher initial investments. For BIRDS, total life-cycle costs are calculated as described above for all building design options for all study periods. The user of the tool has the option to select any of the building designs as the "base case," and compare it to any of the alternative designs. For an investor comparing mutually exclusive design alternatives, the same study period must be used for all alternatives. For those interested in the sensitivity of LCC results to the assumed study period length, BIRDS permits the study period length for a given building design to vary.

Two metrics are used to analyze changes in life-cycle costs: net LCC savings and net LCC savings as a percentage of base case LCC. Net LCC savings (NS) is the difference between the base case LCC (C_{Base}) and alternative design LCC (C_{Alt}) as shown in the following equation:

$$NS = C_{Base} - C_{Alt}$$

Net LCC savings as a percentage of base case LCC (PNS) is the net LCC savings divided by the base case LCC. This metric, shown in the equation below, allows for comparisons across building types that vary significantly in terms of floor area.

$$PNS = \frac{NS}{C_{Base}} * 100$$

6 BIRDS Tutorial

BIRDS is a web-application designed for sustainability performance (energy, cost, and environmental impacts) comparisons for eleven different U.S. commercial building types. BIRDS provides the framework to take an initial building design and make comparisons across different locations, energy standard editions, and/or study periods.

The comparison process has four steps:

1. Select the building prototype to evaluate.
2. Select baseline values and alternatives for comparison (location, standard edition, and study period).
3. Select baseline and alternative weighting preferences for environmental performance.
4. View results graph and data.

BIRDS begins on the "BIRDS-Overview" tab shown in Figure 6-1, which describes the basic information about the purpose of the tool. Once a user is familiar with the purpose of the tool, the user can begin the evaluation process.

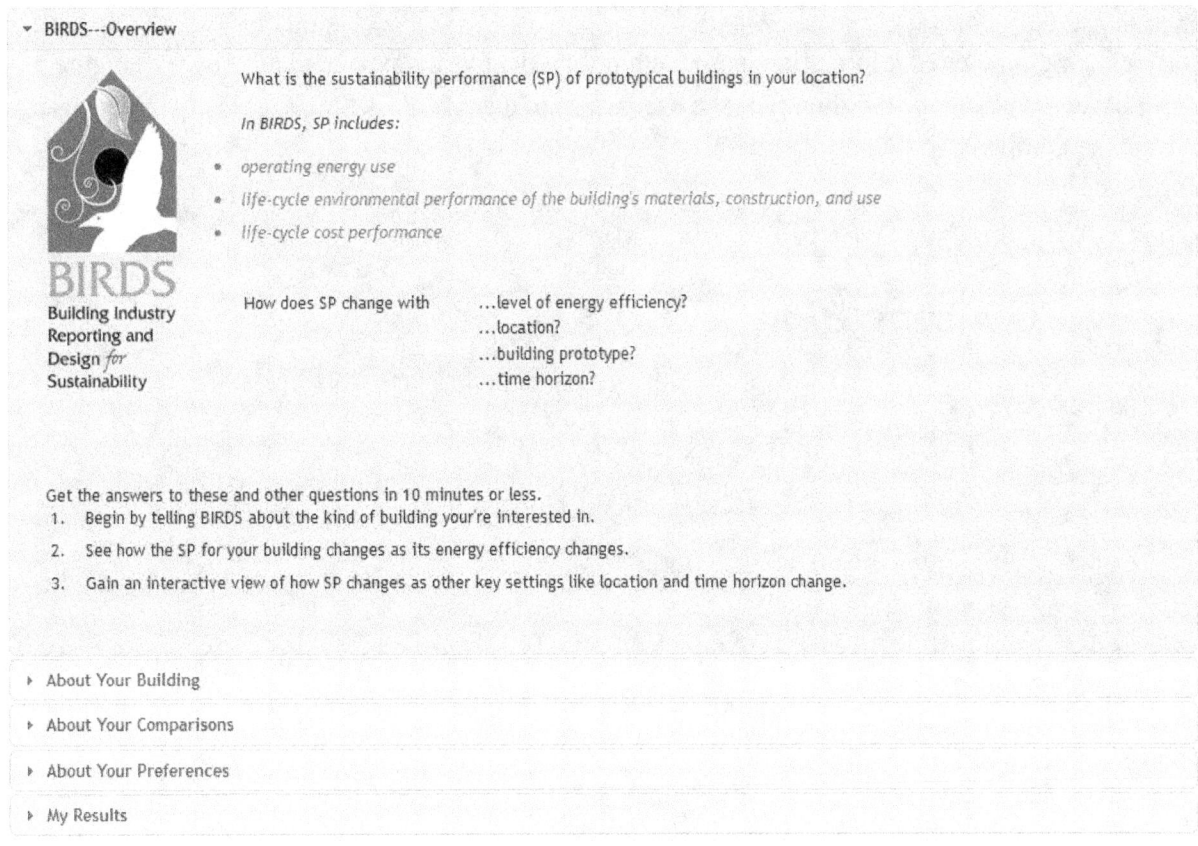

Figure 6-1 BIRDS Overview

6.1 Selecting Building Prototype

From the "BIRDS – Overview" tab the user clicks on the "About Your Building" tab. The section expands, as shown in Figure 6-2. The user selects the My Building Prototype information by selecting Building Type and Number of Floors from the dropdown menus. Note that only apartments, dormitories, and office buildings have more than one option for the number of floors. The remaining building types have a default number of floors.

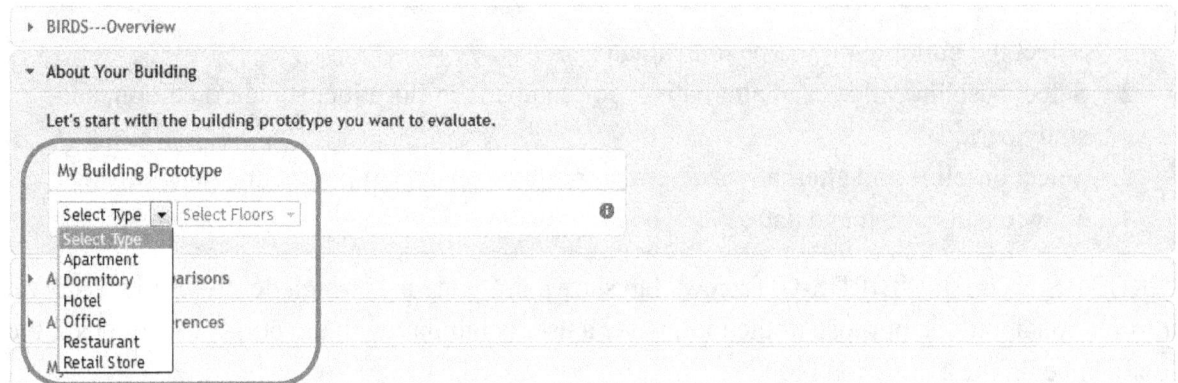

Figure 6-2 Selecting Building Prototype

By clicking on the red information icon, the building details are expanded as shown in Figure 6-3. After selecting the building type the user wants to analyze, it is time to select what the user wants to compare.

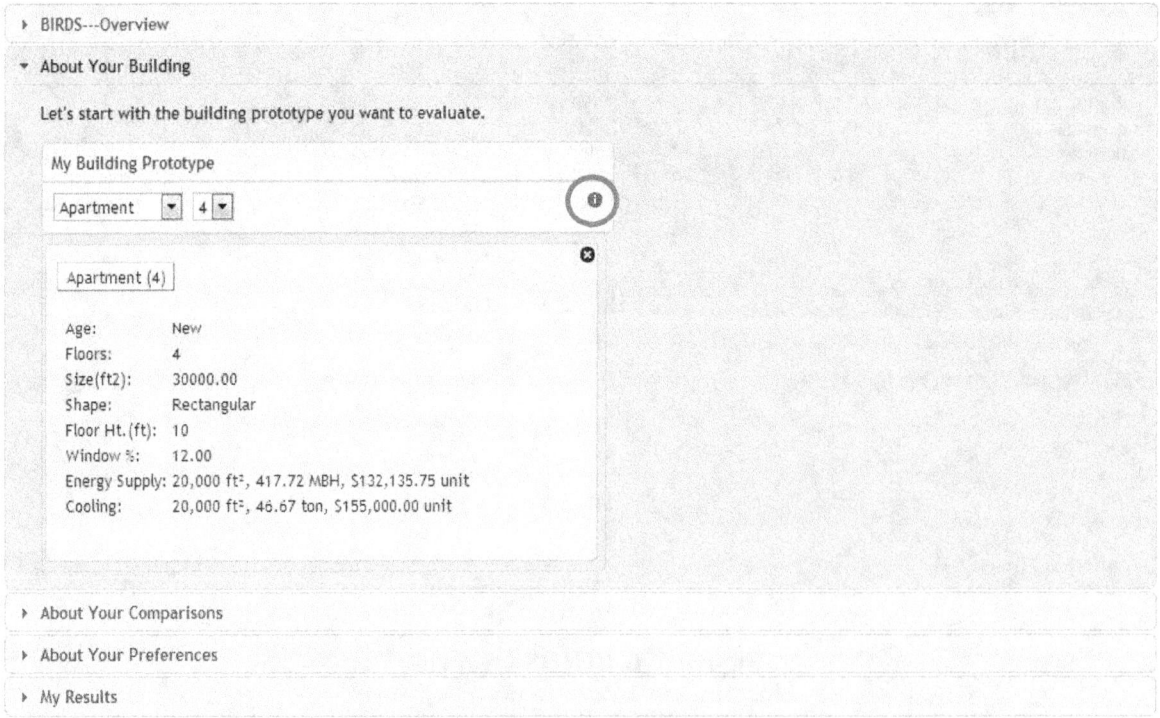

Figure 6-3 Building Prototype Details

6.2 Selecting Comparisons

Clicking on the "About Your Comparisons" tab displays dropdown menus for the preferred Baseline Values for the building's State, City, Standard Edition, and Study Period as shown in Figure 6-4. These are the baseline values that will be used for all comparisons. Note that all baseline values must be defined or an error will occur in the results. For illustration purposes, the Baseline Values are Anchorage, Alaska, using *ASHRAE 90.1-1999* and a 10-year study period.

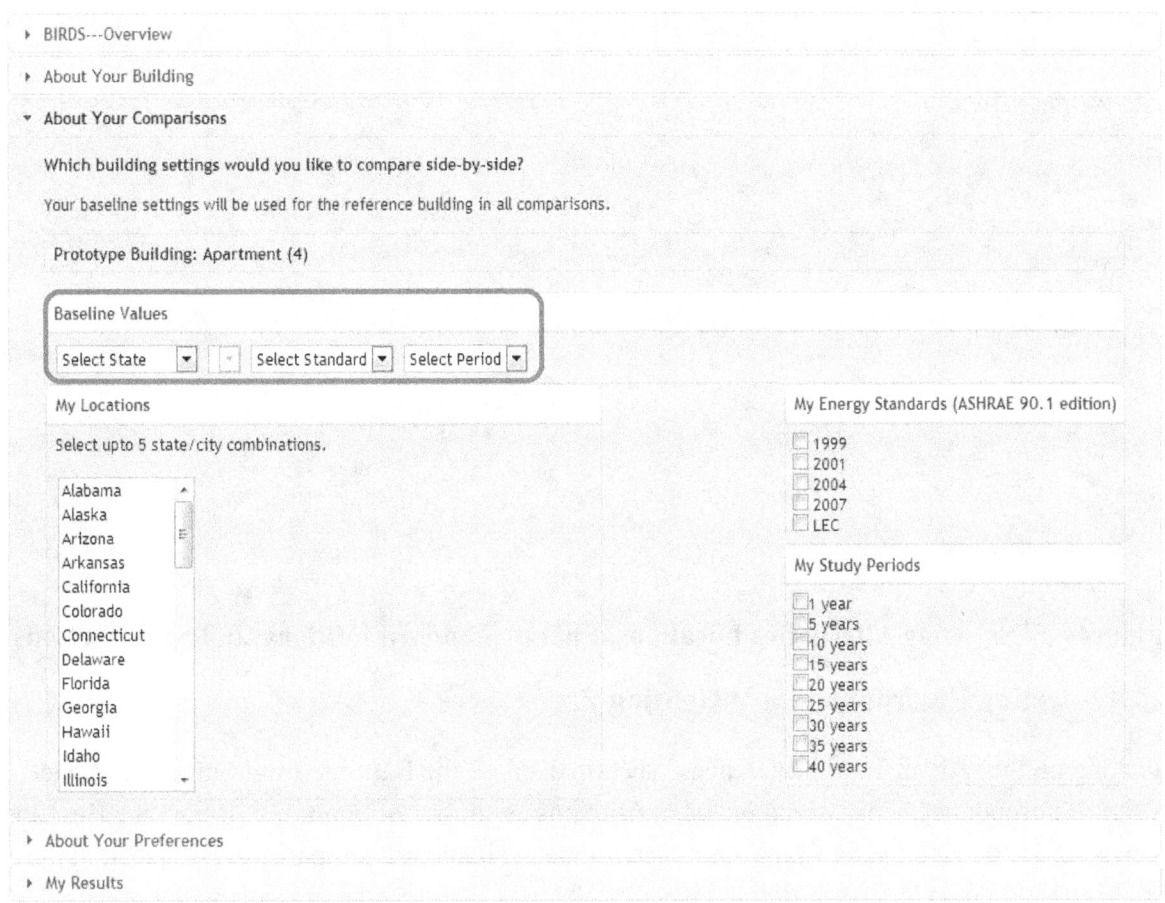

Figure 6-4 Selecting Comparisons

Next, the user can select alternative locations for comparison. After selecting a state from the scroll box, a checkbox list of available cities in that state will appear. The city/state combination will be added to the list to the right, which can contain up to 5 locations for comparison, as shown in Figure 6-5. The user can select up to 5 alternative Energy Standard editions from the checkbox list at the top right, which include *ASHRAE Standards 90.1-1999*, *90.1-2001*, *90.1-2004*, *90.1-2007*, and a Low Energy Case (LEC) based on *ASHRAE 189.1-2009*. Finally, the user can select up to 9 alternative Study Periods in the checkbox list at the bottom right, which range from 1 year to 40 years. Once the user has defined the baseline values and the alternative values for comparison, the user needs to define environmental weighting preferences.

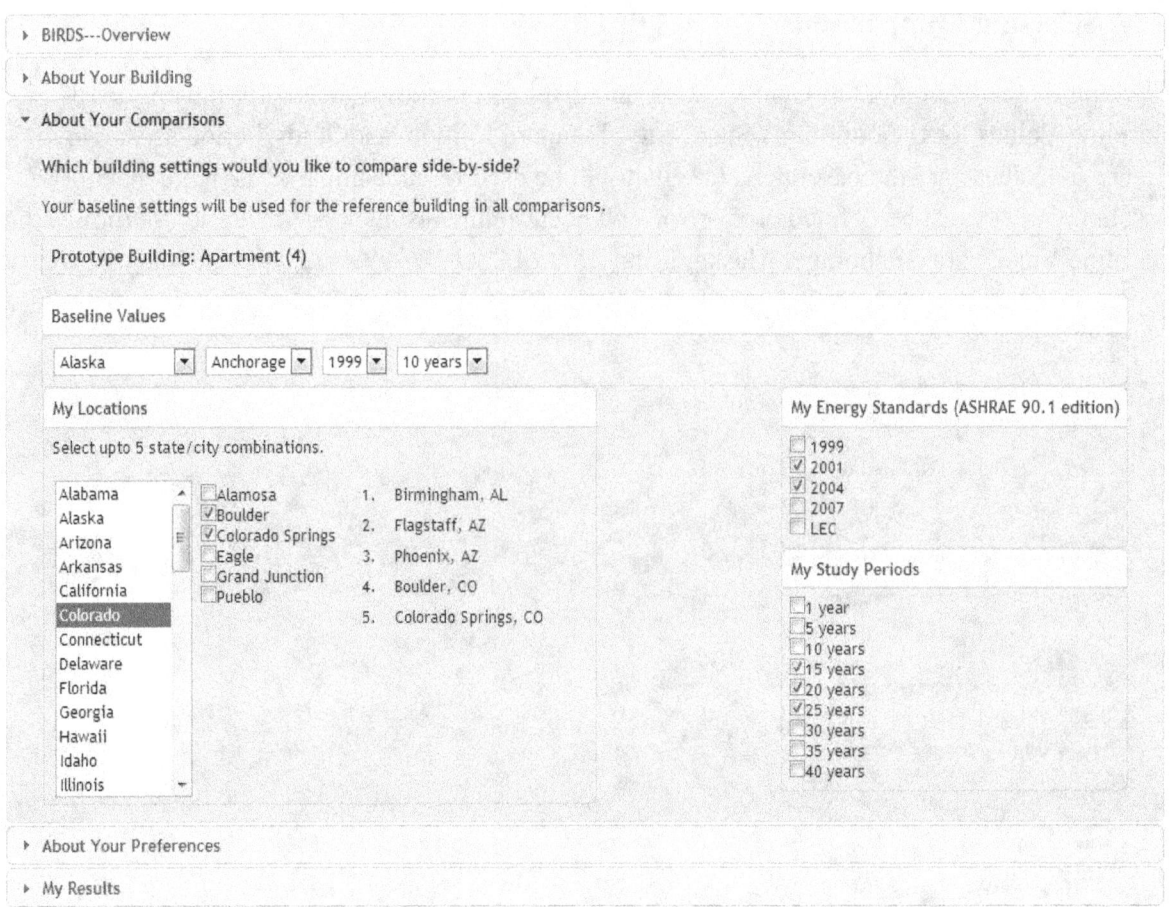

Figure 6-5 Selecting Alternative Locations, Energy Standard Editions, and Study Periods

6.3 Selecting Environmental Weighting Preferences

Clicking on the "About Your Preferences" section displays the Baseline Environmental Impact Weightset dropdown menu, which includes 5 options as shown in Figure 6-6. Below the Baseline Weightset are the alternative weight-sets that will be available for comparison. By selecting the red information icon in the Pre-defined weights, the user can view the environmental impact values for pre-defined weight-sets.

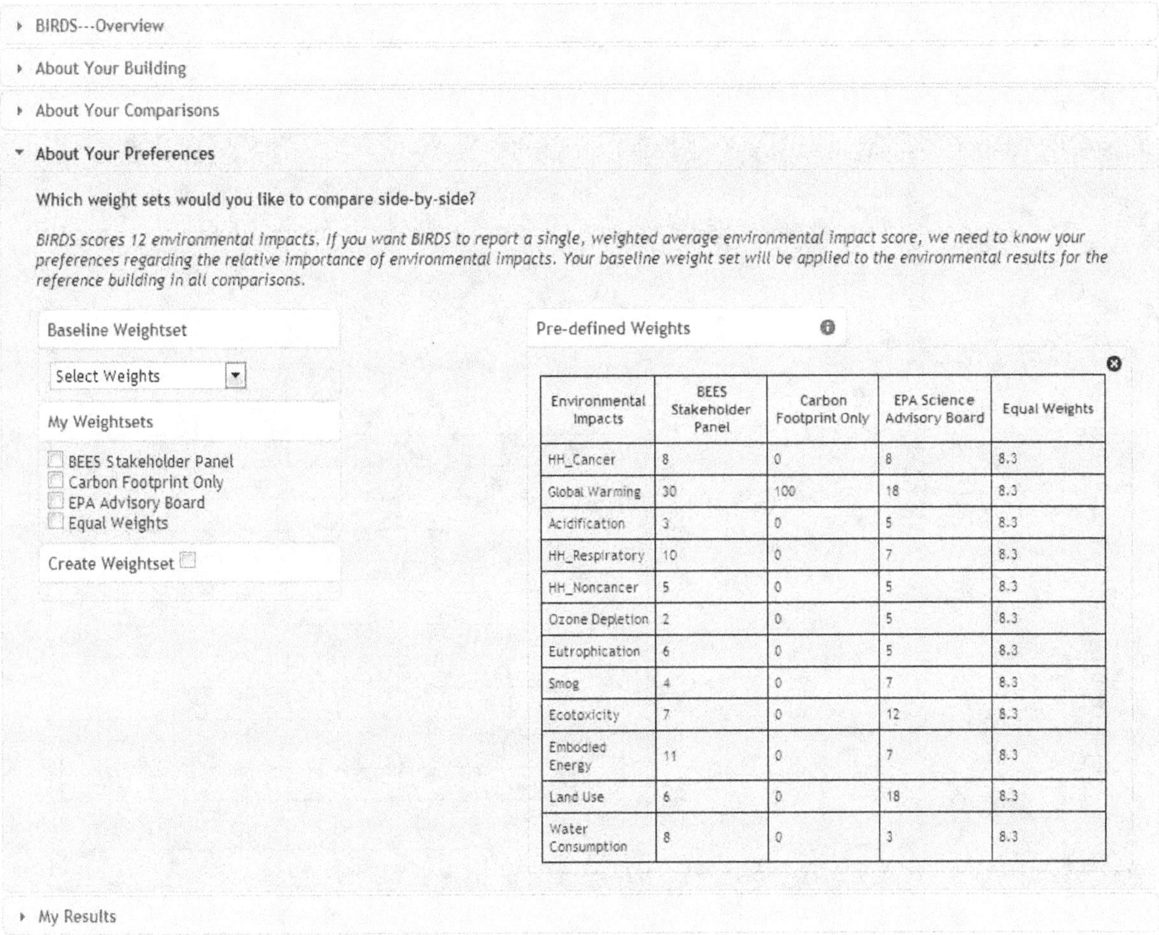

Which weight sets would you like to compare side-by-side?

BIRDS scores 12 environmental impacts. If you want BIRDS to report a single, weighted average environmental impact score, we need to know your preferences regarding the relative importance of environmental impacts. Your baseline weight set will be applied to the environmental results for the reference building in all comparisons.

Baseline Weightset

Select Weights ▾

My Weightsets

☐ BEES Stakeholder Panel
☐ Carbon Footprint Only
☐ EPA Advisory Board
☐ Equal Weights

Create Weightset ☐

Pre-defined Weights   ⓘ   ✖

Environmental Impacts	BEES Stakeholder Panel	Carbon Footprint Only	EPA Science Advisory Board	Equal Weights
HH_Cancer	8	0	8	8.3
Global Warming	30	100	18	8.3
Acidification	3	0	5	8.3
HH_Respiratory	10	0	7	8.3
HH_Noncancer	5	0	5	8.3
Ozone Depletion	2	0	5	8.3
Eutrophication	6	0	5	8.3
Smog	4	0	7	8.3
Ecotoxicity	7	0	12	8.3
Embodied Energy	11	0	7	8.3
Land Use	6	0	18	8.3
Water Consumption	8	0	3	8.3

▸ My Results

Figure 6-6 Selecting Environmental Weighting Preferences

The user is given flexibility to create a custom weight-set by checking the Create Weightset checkbox, which can be used as the baseline or as an alternative. As shown in Figure 6-7, checking the box brings up a list of the 12 environmental impact categories. Each category must be given a weight between 0 and 100, with the sum of all 12 weights adding up to 100. Once a custom weight-set is defined it will become a selection available in the Baseline Weightset dropdown and as a checkbox in the My Weightsets alternative options. At this point, all the necessary user inputs have been defined and the user can now look at the results.

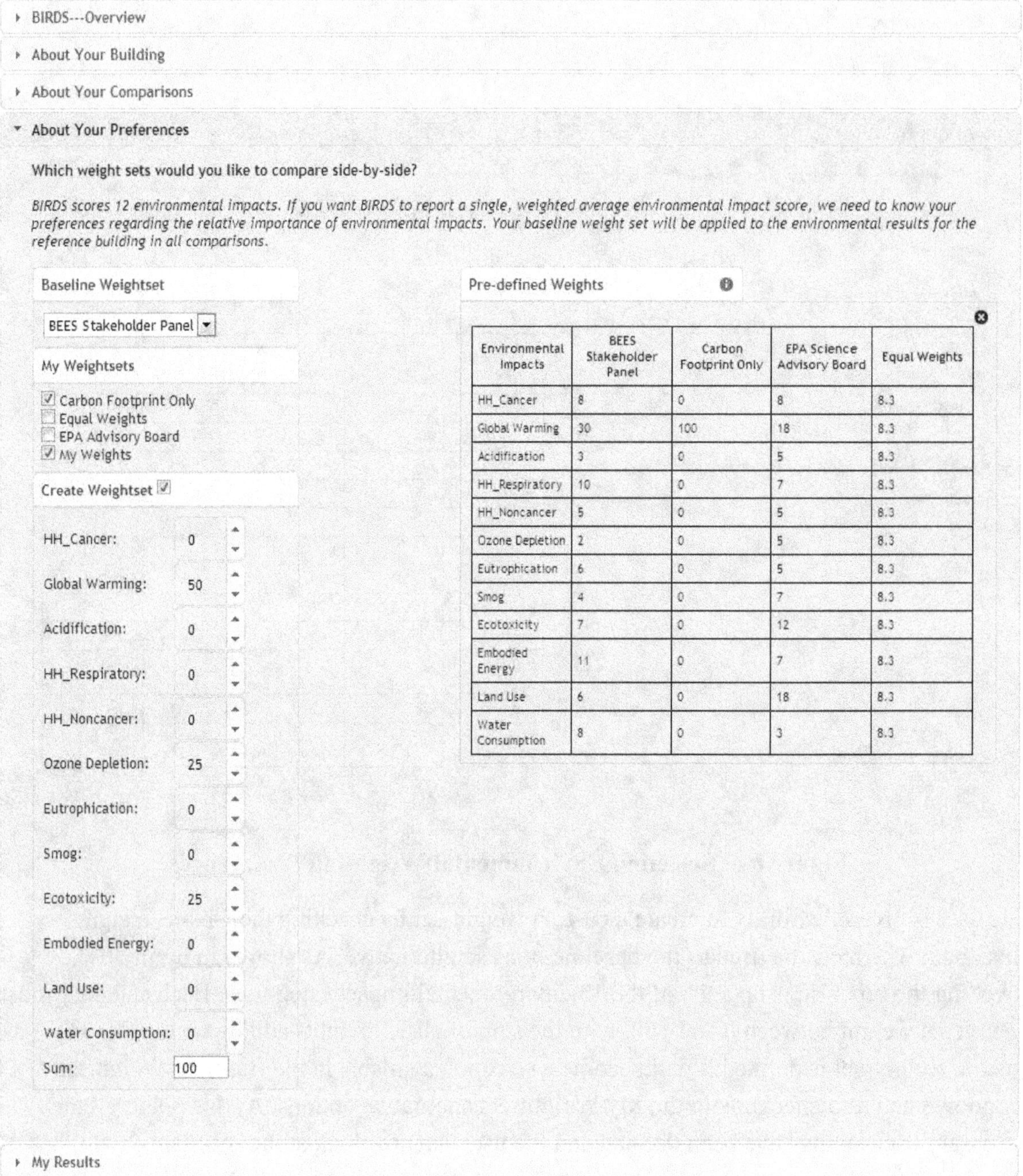

▸ BIRDS---Overview

▸ About Your Building

▸ About Your Comparisons

▾ About Your Preferences

Which weight sets would you like to compare side-by-side?

BIRDS scores 12 environmental impacts. If you want BIRDS to report a single, weighted average environmental impact score, we need to know your preferences regarding the relative importance of environmental impacts. Your baseline weight set will be applied to the environmental results for the reference building in all comparisons.

Baseline Weightset
BEES Stakeholder Panel ▾

My Weightsets

☑ Carbon Footprint Only
☐ Equal Weights
☐ EPA Advisory Board
☑ My Weights

Create Weightset ☑

HH_Cancer:	0	▲▼
Global Warming:	50	▲▼
Acidification:	0	▲▼
HH_Respiratory:	0	▲▼
HH_Noncancer:	0	▲▼
Ozone Depletion:	25	▲▼
Eutrophication:	0	▲▼
Smog:	0	▲▼
Ecotoxicity:	25	▲▼
Embodied Energy:	0	▲▼
Land Use:	0	▲▼
Water Consumption:	0	▲▼
Sum:	100	

Pre-defined Weights ⓘ

Environmental Impacts	BEES Stakeholder Panel	Carbon Footprint Only	EPA Science Advisory Board	Equal Weights
HH_Cancer	8	0	8	8.3
Global Warming	30	100	18	8.3
Acidification	3	0	5	8.3
HH_Respiratory	10	0	7	8.3
HH_Noncancer	5	0	5	8.3
Ozone Depletion	2	0	5	8.3
Eutrophication	6	0	5	8.3
Smog	4	0	7	8.3
Ecotoxicity	7	0	12	8.3
Embodied Energy	11	0	7	8.3
Land Use	6	0	18	8.3
Water Consumption	8	0	3	8.3

▸ My Results

Figure 6-7 Defining a Custom Weightset

6.4   Viewing Results

Clicking on the "My Results" tab will display the "Select Chart Options" section. Three different chart types are available in the application: Life-Cycle Cost, Operating Energy, or Environmental Impact Score (EIS). As shown in Figure 6-8, the user must select the Chart Type from the dropdown menu, the Baseline for the comparison to be made, and the Units in which the user prefers the results. Note that the units include a per unit of floor area impact, which is only reported in square feet and not square meters because the tool is designed for use domestically, which predominantly uses I-P unit instead of metric units.

Figure 6-8 Selecting Chart Options

Once the user has made these selections, the user presses the "View Graph" button. The graph with corresponding data table is displayed. Figure 6-9 shows an example of the Life-Cycle Cost results with a graph of the total life-cycle costs in present value dollars for a 4-story apartment building built to meet *ASHRAE 90.1-1999* across the different study periods selected by the user

(10, 15, 20, and 25 years). As can be seen in the graph, total present value life-cycle costs increase as the study period increases in length, which is a result of additional operational energy costs and maintenance, repair, and replacement costs during those additional years.

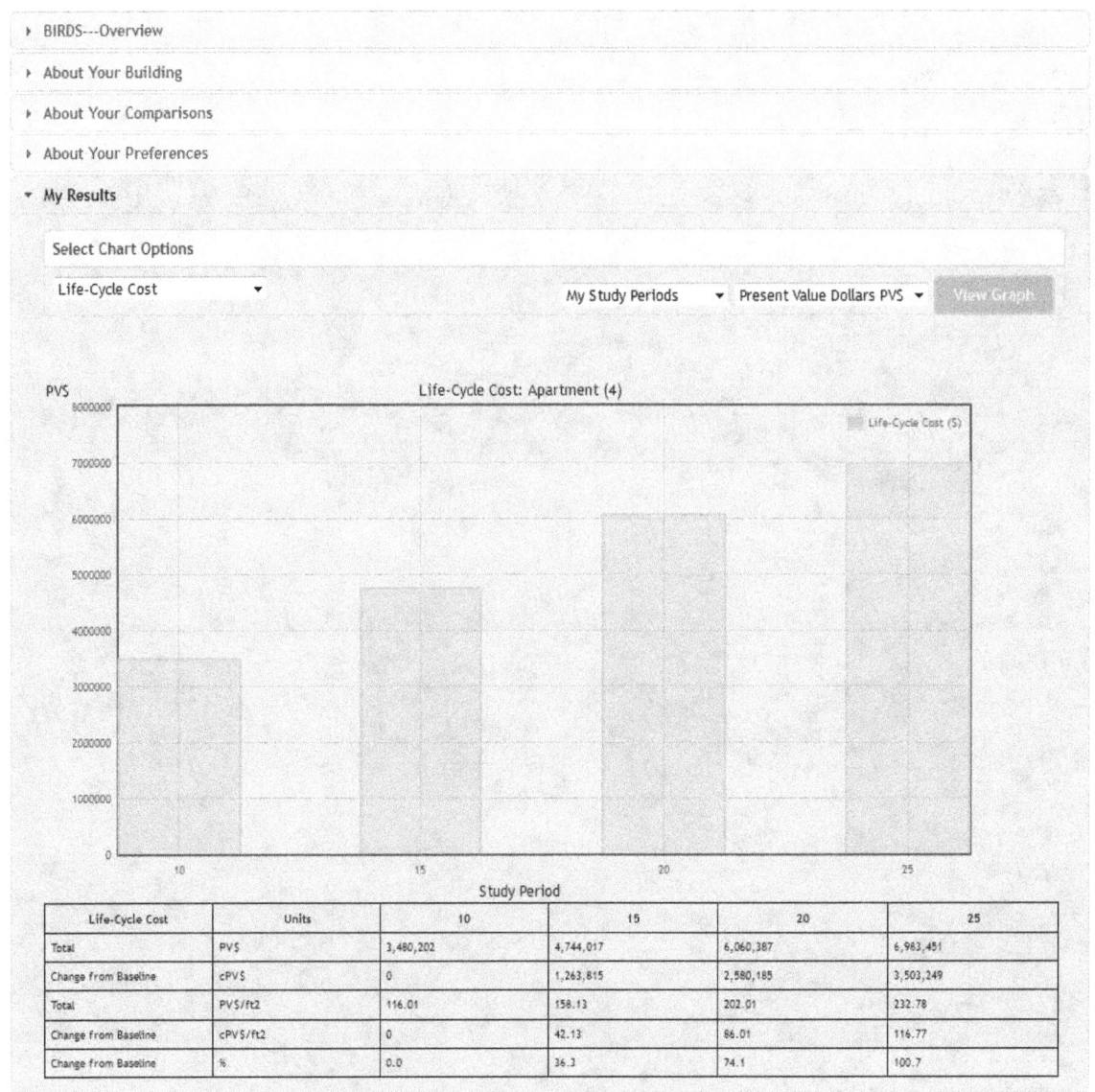

Figure 6-9 Life-Cycle Cost Graph

The data table is more comprehensive, and includes all of the potential metrics available for comparisons across study periods for the baseline location, which include total life-cycle costs, change in total life-cycle costs relative to the 10-year baseline, total life-cycle costs per square foot of floor area, change in total life-cycle costs from the baseline per square foot of floor area, and percentage change in life-cycle costs relative to the 10-year baseline.

Figure 6-10 shows an example of the Operating Energy results with a graph of the change in annual energy consumption per square foot of floor area for a 4-story apartment building built to

126

meet *ASHRAE 90.1-1999* across the different locations selected by the user, including cities in Alabama, Alaska, Arizona, and Colorado. As can be seen in the graph, Anchorage, Alaska consumes more energy per unit of floor area than any of the other cities considered in the analysis. Colorado Springs consumes 5 kBtu/year/ft^2 less energy than the same building in Anchorage.

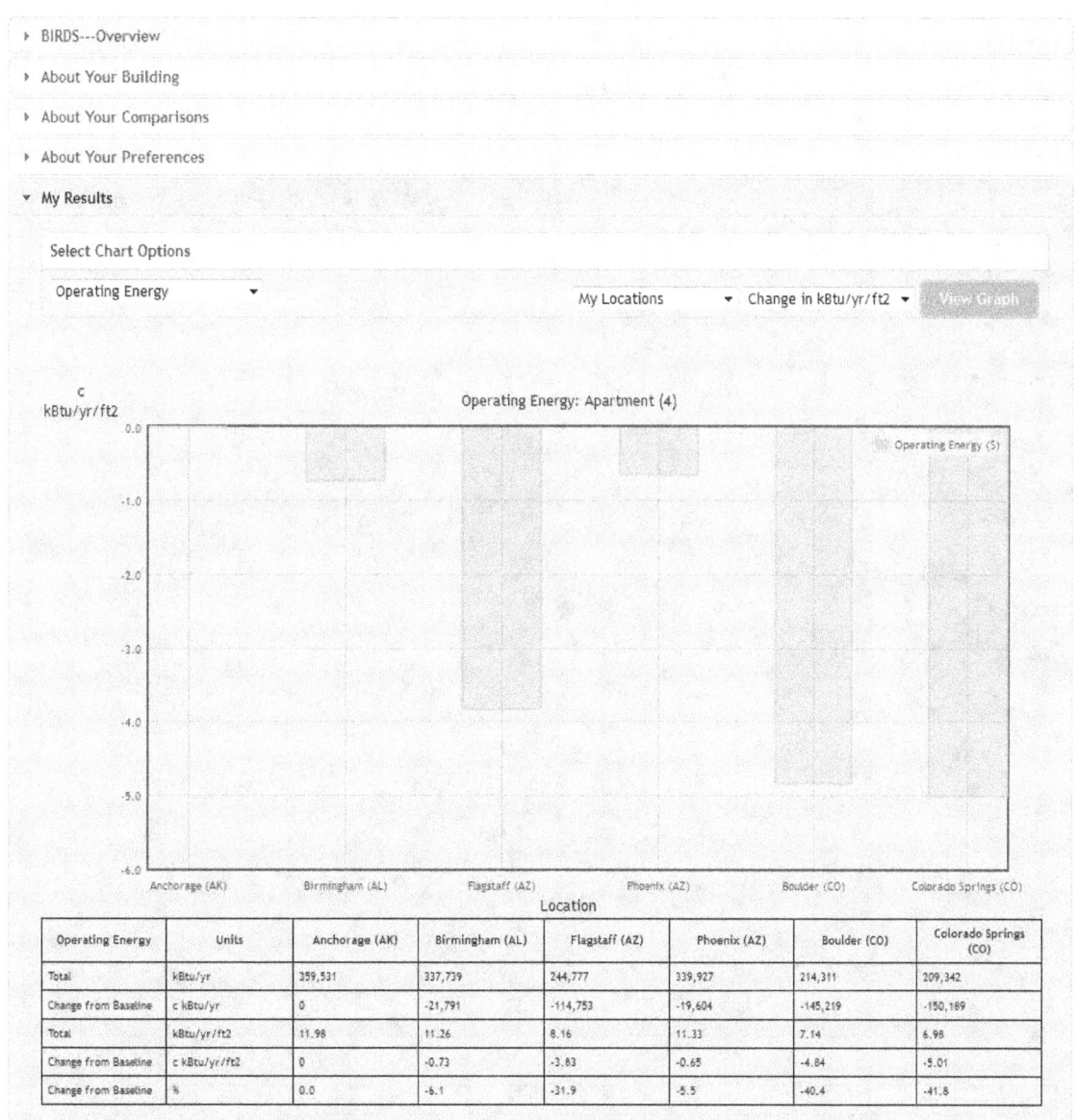

Figure 6-10 Operating Energy Consumption Graph

The data table shows all the metrics a user can use for comparisons: total annual energy consumption, change in total annual energy consumption, total annual energy consumption per square foot of floor area, change in total annual energy consumption per square foot of floor area, and percentage change in energy consumption relative to the baseline.

When the EIS Chart is selected an additional drop down menu is displayed that allows the user to select either the EIS or the total flows for one of the 12 environmental impact categories. Figure 6-11 shows a graph of the total global warming potential impacts in kilograms of CO_2e emissions. The newest edition of *ASHRAE 90.1* considered in the analysis (*ASHRAE 90.1-2004*) realizes the lowest impact on global warming potential. The data table shows all the metrics a user can use for comparisons: total CO_2e flows, change in total CO_2e flows, total CO_2e flows per square foot of floor area, change in total CO_2e flows per square foot of floor area, and percentage change in total flows relative to the total flows for the baseline.

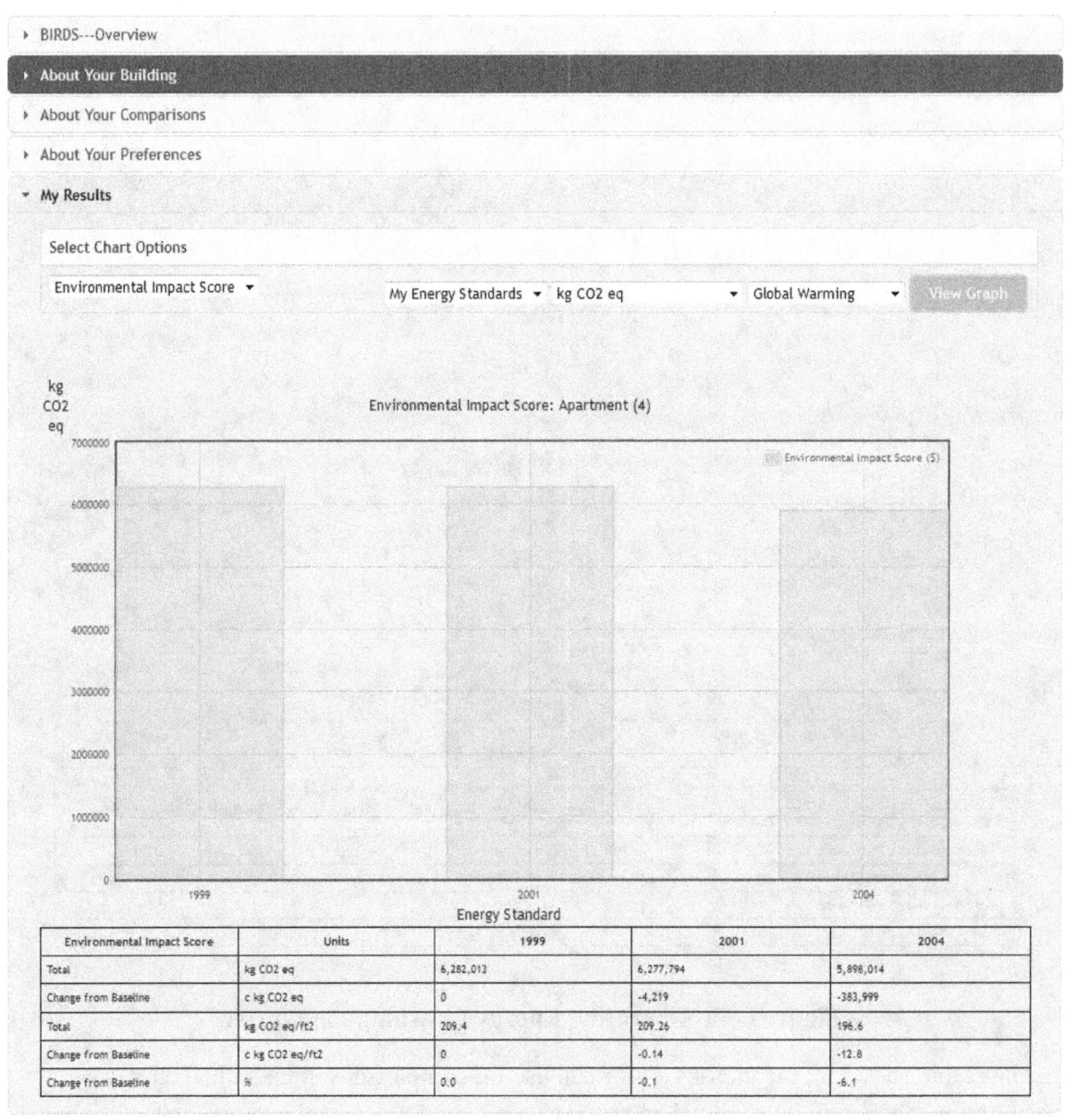

Environmental Impact Score	Units	1999	2001	2004
Total	kg CO2 eq	6,282,013	6,277,794	5,898,014
Change from Baseline	c kg CO2 eq	0	-4,219	-383,999
Total	kg CO2 eq/ft2	209.4	209.26	196.6
Change from Baseline	c kg CO2 eq/ft2	0	-0.14	-12.8
Change from Baseline	%	0.0	-0.1	-6.1

Figure 6-11 Environmental Impact – Global Warming Potential Graph

7   Limitations and Future Research

BIRDS is limited in scope and would be strengthened by including uncertainty analysis, expanding the database and metrics, and adding design flexibility to the tool.

Uncertainty analysis is needed for at least three elements of the analysis. First, consider the assumed discount rate. Although 3 % is a reasonable discount rate, in real terms, for federal government investment decisions, it may be too low of a value for an expected real return on an alternative investment in the private sector. Sensitivity analysis on the assumed discount rate is needed to determine the robustness of the cost results. Second, the current analysis assumes that building cooling loads are met by equipment running on electricity while heating loads are met by equipment running on natural gas, which is not the typical fuel mix for some areas of the nation. The database should be expanded to include alternative fuel source options, such as heating oil use in the New England area. Third, the BIRDS environmental impact scores do not incorporate uncertainty analysis as required by international standards (ISO, 2006). While incorporating uncertainty analysis is problematic due to a lack of underlying uncertainty data, this omission should be brought into the interpretation of the BIRDS results.

Additional data are needed to refine and expand the BIRDS database. The 11 prototypical buildings analyzed in this study may not be representative of the entire building stock for each building type. For example, all high-rise buildings are not 100 % glazed. For this reason, the results should be considered as general magnitudes for making reasonable comparisons instead of hard numbers. Future research should include additional prototypes, such as the DOE Benchmark Buildings, in the database. Additionally, since existing buildings account for nearly the entire building stock, prototypes for energy retrofits to buildings should be incorporated into the BIRDS database as well. The state average energy cost rates and energy-related carbon emissions rates do not control for local variation in energy tariffs or electricity fuel mixes. By using utility-level energy cost and emissions rate data, the accuracy of the estimates in BIRDS could be improved.

The analysis in this study ignores the impacts that plug and process loads have on the reductions in energy use. Buildings with greater plug and process loads will realize smaller percentage changes in energy use because the energy efficiency measures considered in this study focus on the building envelope and HVAC equipment, holding constant the energy use from other equipment used in the building. As building energy efficiency improves, the plug and process loads become a larger fraction of the overall energy load. Future research should consider the impact the assumed plug and process loads have on the overall energy savings realized by energy efficiency improvements to buildings.

Properly interpreting the BIRDS environmental performance results requires placing them in perspective. The environmental impact scores assess the life-cycle impacts of operating energy use based on inventories of localized energy simulation results and regional electricity grids. All other elements of the scores—including a building's use of materials and its water consumption over the study period—are based on U.S. average life-cycle inventory data for prototypical buildings. The baseline data for these buildings represent status quo building technologies as of 2002, the year of the latest available input-output data from the U.S. Bureau of Economic Analysis. To account for evolution in status quo technologies over time, future versions of BIRDS should incorporate newer releases of these data as they become available.

The BIRDS results do not apply to buildings constructed in other countries where industry practices, fuel mixes, environmental regulations, transportation distances, and labor and material markets may differ. Furthermore, all buildings of a given type are not created equal. Building designs, sizes, useful lives, materials compositions, and costs will all vary for an individual building. The BIRDS results for a building prototype do not necessarily represent the performance of an individual building of that type. Future versions of the tool should permit flexibility in building design and use of materials.

The BIRDS LCAs use selected inventory flows converted to selected local, regional, and global environmental impacts to assess environmental performance. Those inventory flows which currently do not have scientifically proven or quantifiable impacts on the environment are excluded, such as mineral extraction and wood harvesting which are qualitatively thought to lead to loss of habitat and an accompanying loss of biodiversity. If the BIRDS user has important knowledge about these issues, it should be brought into the interpretation of the BIRDS results.

The Environmental Problems approach that BIRDS uses for impact assessment does not offer the same degree of relevance for all environmental impacts. For global and regional effects (e.g., global warming and acidification) the method may result in an accurate description of the potential impact. For impacts dependent upon local conditions (e.g., smog, ecological toxicity, and human health impacts) it may result in an oversimplification of the actual impacts because the indices are not tailored to localities.

Life cycle impact assessment is a rapidly evolving science. Assessment methods unheard of a decade ago have since been developed and are now being used routinely in LCAs. While BIRDS incorporates state-of-the-art impact assessment methods, the science will continue to evolve and methods in use today—particularly those for land and water use—are likely to change and improve over time. Future versions of BIRDS should incorporate these improved methods as they become available.

During the interpretation step of the BIRDS LCAs, environmental impact results are optionally combined into a single environmental performance score using relative importance weights. These weights necessarily incorporate values and subjectivity. BIRDS users should routinely test the effects on the environmental impact scores of changes in the set of importance weights.

Energy, environmental, and economic performance are but three attributes of building performance. The BIRDS model assumes that its building prototypes all meet minimum technical performance requirements. However, there may be significant differences in technical performance not evaluated in BIRDS, such as acoustic or fire performance, which may outweigh energy, environmental, and economic considerations.

References

Ander, G.D. (2012). *Daylighting*, written for the Whole Building Design Guide, a program of the National Institute of Building Sciences. http://www.wbdg.org/resources/daylighting.php.

ASHRAE/IESNA Standard Project Committee 90.1, ASHRAE 90.1-1999 Standard- Energy Standard for Buildings Except Low-Rise Residential Buildings, 1999, ASHRAE, Inc.

ASHRAE/IESNA Standard Project Committee 90.1, ASHRAE 90.1-2001 Standard- Energy Standard for Buildings Except Low-Rise Residential Buildings, 2001, ASHRAE, Inc.

ASHRAE/IESNA Standard Project Committee 90.1, ASHRAE 90.1-2004 Standard- Energy Standard for Buildings Except Low-Rise Residential Buildings, 2004, ASHRAE, Inc.

ASHRAE/IESNA Standard Project Committee 90.1, ASHRAE 90.1-2007 Standard- Energy Standard for Buildings Except Low-Rise Residential Buildings, 2007, ASHRAE, Inc.

ASTM International (2011), Standard Practice for Applying the Analytic Hierarchy Process to Multiattribute Decision Analysis of Investments Related to Buildings and Building Systems, ASTM Designation E1765-11, West Conshohocken, PA.

ASTM International (2010), Standard Practice for Measuring Life-Cycle Costs of Buildings and Building Systems, ASTM Designation E917-05, West Conshohocken, PA.

ASTM International (2012), Standard Guide for Selecting Economic Methods for Evaluating Investments in Buildings and Building Systems, ASTM Designation E1185-12, West Conshohocken, PA.

Atherton, J., International Council on Mining & Metals (ICMM) (2006). Declaration by the Metals Industry on Recycling Principles, *International Journal of Life Cycle Assessment*, January 2007, Volume 12, Issue 1, pp 59-60.

Atlas Roofing Corporation (2008). MSDS - Glass Reinforced Facer (GRF).

Atlas Roofing Corporation (2012). MSDS - Energy Shield/Energy Shield Plus/Cavity Wall.

Bakane, T. (2009). Properly Diagnosing Chiller Life Cycles. http://www.facilitiesnet.com/hvac/article/Properly-Diagnosing-Chiller-Life-Cycles--10645#

Baltimore Aircoil Company (BAC) *Product and Application Handbook –Series 1500 Engineering Data on Cooling Tower* (no date). Accessed in May 2012 from http://www.baltimoreaircoil.com/english/resource-library/file/543

Baltimore Aircoil Company (BAC) (2010). MS Word Document entitled *"Series 1500 Cooling Tower Specification"* (no date), Accessed in May 2012 from http://www.baltimoreaircoil.com/english/products/cooling-towers/series-1500/specifications.

Bare, J. (2011), "TRACI 2.0: The Tool for the Reduction and Assessment of Chemical and Other Environmental Impacts 2.0," *Clean Technologies and Environmental Policy,* published online January 2011.

Bare, J. et al. (2006) "Development of the Method and U.S. Normalization Database for Life Cycle Impact Assessment and Sustainability Metrics," *Environmental Science and Technology,* 40: 16, 5108-5115.

Belzer, D.B., et al. (2005). *Analysis of Potential Benefits and Costs of Adopting a Commercial Building Energy Standard in South Dakota*, Pacific Northwest National Laboratory, PNNL-15101.

Biswas, W. and M. Rosano, Curtin University (Perth) (2011). "A life cycle greenhouse gas assessment of remanufactured refrigeration and air conditioning compressors", *Int. J. Sustainable Manufacturing*, Vol. 2, Nos. 2/3.

Brundtland Commision, *Our Common Future: Report of the UN Commission on Environment and Development*, Oxford University Press, 1987.

Dewhurst, D., Smith Mechanical Quantity Survey Ltd., Jan 2012 data compilation.

Energy Independence and Security Act of 2007, P.L. 110-140, Sec 441.

EUMEPS (2002). *Behaviour of EPS in case of fire*. Retrieved July 2012 at http://www.eumeps.org/show.php?ID=4471&psid=hmotjteo.

Extruded Polystyrene Foam Association (XPSA) website. http://www.xpsa.com/.

Flakeboard, no date. Product Data Sheet: FIBREX thin high density fiberboard. http://www.flakeboard.com.

Franklin Associates, A Division of ERG (Eastern Research Group) (2009). Energy and Greenhouse Gas Savings for EPS Roam Insulation Applied to Exterior Walls of Single Family Residential Housing in the U.S. and Canada, prepared for the EPS Molders Association.

Fuller, S. and Petersen, S. (1996) *Life-Cycle Costing Manual for the Federal Energy Management Program*, NIST Handbook 135, 1995 Edition.

GAF Materials Corporation (2010). EnergyGuard Perlite Roof Insulation Data Sheet.

GAF Materials Corporation (2008). EnergyGuard MSDS #2060.

Georgia Pacific (2009). Material Safety Data Sheet: Glass Mat-Faced Gypsum Panels.

Gloria, T. P., Lippiatt, B. C., and Cooper J. (2007) "Life Cycle Impact Assessment Weights to Support Environmentally Preferable Purchasing in the United States," *Environmental Science and Technology,* 41, 7551-7557.

Goedkoop, M.J. et al., *Recipe 2008, A Life Cycle Impact Assessment Method which comprises Harmonized Category Indicators at the Midpoint and the Endpoint Level; First edition Report I: Characterisation;* http://www.lcia-recipe.net/, 2009.

Greenheck, Engineering Update (Summer 2004), "Gauge and Weight Chart for Sheet Steel, Galvanized Steel, Stainless Steel, and Aluminum."

Guinee, J.B, et al., *Handbook on Life Cycle Assessment: Operational Guide to the ISO Standards*, Kluwer Academic Publisher, 2002.

Halverson, M.A., Gowri, K., and Richman, E.E. (2006). *Analysis of Energy Savings Impacts of New Commercial Energy Codes for the Gulf Coast*, Pacific Northwest National Laboratory, PNNL-16282.

Hendrickson, C. T., Lave, L. B., and Matthews, H. S. *Environmental Life Cycle Assessment of Goods and Services: An Input-Output Approach*, Resources for the Future Press, 2006.

Institute for Construction and Environment (2011), EPD according to ISO 14025: *Trane Centrifugal Chiller 2500 Ton*, EPD-TRA-2011111-E. http://www.bau-umwelt.com.

Intergovernmental Panel on Climate Change (IPCC) and the Montreal Protocol's Technology and Economic Assessment Panel (2005). IPCC/TEAP Special Report: "Safeguarding the Ozone Layer and the Global Climate System."

International Copper Association (ICA) (2012). Life Cycle Assessment of Primary Copper Cathode.

International Organization for Standardization (ISO), *Environmental Management--Life-Cycle Assessment--Principles and Framework*, International Standard 14040, 2006.

International Organization for Standardization (ISO), Environmental Management – Life Cycle Assessment – Requirements and Guidelines, International Standard 14044, 2006.

Johns Manville (2010). Fesco Roof Insulation Safety Data Sheet ID 3002.

Kawneer, an Alcoa Company (2011). Versoleil SunShade - Outrigger System, No. EC 97911-34.

Kneifel, J. (2010), "Life-cycle Carbon and Cost Analysis of Energy Efficiency Measures in New Commercial Buildings," *Energy and Buildings* 42: 3, 333-340.

Kneifel, J. (2011a), "Beyond the Code: Energy, Carbon, and Cost Savings using Conventional Technologies," *Energy and Buildings*, 43: 951-959.

Kneifel, J. (2011b). *Prototype Commercial Buildings for Energy and Sustainability Assessment: Whole Building Energy Simulation Design*, NIST Technical Note 1716.

Kneifel, J. (2012). *Prototype Commercial Buildings for Energy and Sustainability Assessment: Design Specification, Life-Cycle Costing and Carbon Assessment*, NIST Technical Note 1732.

LCAFood database, found at http://www.lcafood.dk/processes/industry/potatoflourproduction.htm.

Legutko, T. and M. Taylor, Carrier Corporation (2000). *Split Systems: A Primer*.

Levin, L. "Best Sustainable Indoor Air Quality Practices in Commercial Buildings," *Third International Green Building Conference and Exposition--1996*, NIST Special Publication 908, Gaithersburg, MD, November 1996, p. 148.

McQuay International Product Catalog 611-1: Air-Cooled Scroll-Compressor Chillers, Model AGZ-D • 25 to 190 Tons • R-410A • 60Hz/50Hz.

Nath, N.G. & D.L. Ruff (1982). Perlite boards and method for making same. Publication No. U.S. 4313997 A. Found at: http://www.google.com/patents/US4313997.

National Renewable Energy Laboratory (NREL) (2005-present). U.S. Life-Cycle Inventory Database, Golden, CO. Found at: http://www.nrel.gov/lci/database.

PE International (2010). *Eco-Profile of Aromatic Polyester Polyols (APP)*, performed for PU Europe, Federation of European rigid Polyurethane Foam Associations.

PE International, GaBi database (2011).

Phelan J. and G. Pavlovich, Bayer MaterialScience (2008) "Energy and Environmental Benefits of Insulating Commercial Buildings with Polyiso." (Phelan, 2008)

Phelan J. and G. Pavlovich (Bayer MaterialScience), and J. Jewell (PE Americas) (2010). *Life Cycle Assessment of Polyiso Insulation – Final Report, prepared for Polyisocyanurate Insulation Manufacturers Association (PIMA)* (Bayer MaterialScience, 2010).

RS Means *CostWorks* databases, accessed 2009, http://www.meanscostworks.com/.

Rushing, A., and Lippiatt, B. (2009) *Energy Price Indices and Discount Factors for Life-Cycle Cost Analysis-April 2009*, NISTIR 85-3273-24.

Salazar, James, University of British Columbia (2007). Life Cycle Assessment Case Study of North American Residential Windows – A Thesis Submitted in Partial Fulfilment of the Requirements for the Degree of Master of Science in the Faculty of Graduate Studies.

Shah, V.P. *et al.*, 2008. Life cycle assessment of residential heating and cooling systems in four regions in the United States. *Energy and Buildings* 40, 503–513.

Slant/Fin Fine/Line 30 catalog, no. 30-10C (November 2004).

Suh, S. (2005) "Developing a Sectoral Environmental Database for Input-Output Analysis: The Comprehensive Environmental Data Archive of the U.S.," *Economic Systems Research*, 17: 4, 449-469.

Suh, S., (2010) "Comprehensive Environmental Data Archive (CEDA)," In: Murray J, Wood R (Eds.) *The Sustainability Practitioner's Guide to Input-Output Analysis*, Common Ground Publishing.

Suh S, et al. (2004) "System boundary selection in life-cycle inventories using hybrid approaches," *Environmental Science & Technology*, 38: 657-664.

Suh S. and Huppes G (2005) "Methods for Life Cycle Inventory of a Product," *Journal of Cleaner Production*, 13:687-697.

Suh, S. and Lippiatt, B. (2012) "Framework for Hybrid Life Cycle Inventory Databases: A Case Study on the Building for Environmental and Economic Sustainability (BEES) Database," *International Journal of Life Cycle Assessment*, 17: 604-612.

Swiss Centre for Life Cycle Inventories (2007). ecoinvent Data v.2.2, Dubendorf. Found at: http://www.ecoinvent.org.

Swiss Centre for Life Cycle Inventories (2007). ecoInvent Data v.2.2 "Oil boiler 100kW/CH/I U", from EcoInvent Report No 5: Life Cycle Inventories of Energy Systems: Results for Current Systems in Switzerland and other UTCE Countries (ecoinvent Report No. 5, 2007).

Trane Architectural Hydronic Wall Fin brochure, FIN-PRC004-EN (October 2001).

Trane Product Catalog: Split System Condensing Units —RAUJ; Remote Chiller Evaporators 20 - 120 Tons — 50/60 Hz — R-410A.

Trane Catalog Series R Helical Rotary Liquid Chillers, 70-250 Tons Model RTWD Water-Cooled (February 2010). No. RLC-PRC029-EN.

U.S. Department of Energy, Building Technologies Program, *EnergyPlus* Energy Simulation Software, 2009, http://apps1.eere.energy.gov/buildings/energyplus/.

U.S. Energy Information Administration, *Electric Power Annual State Data Tables*, accessed 2009 and 2010, http://www.eia.doe.gov/fuelelectric.html.

U.S. Energy Information Administration, *Natural Gas Navigator*, accessed 2009 and 2010, http://www.eia.doe.gov/oil_gas/natural_gas/info_glance/natural_gas.html.

U.S. EPA Science Advisory Board, *Toward Integrated Environmental Decision-Making*, EPA-SAB-EC-00-011, Washington, D.C., August 2000.

U.S. EPA Science Advisory Board, *Reducing Risk: Setting Priorities and Strategies for Environmental Protection*, SAB-EC-90-021, Washington, D.C., September 1990.

United Nations Environment Programme (UNEP) Report of the Technology and Economic Assessment Panel (TEAP) (2005). *Volume 3: Report of the Task Force on Foam End-of-Life Issues.*

Veissmann Vitorond 200 Triple-pass hot water heating boiler 1255 to 4387 MBH / 368 to 1285 kW Technical Data Manual, No. 5285 428 v1.6. (July 2010).

Vigener, N. & M.A. Brown (2012). *Building Envelope Design Guide – Windows*, written for the Whole Building Design Guide, a program of the National Institute of Building Sciences. http://www.wbdg.org/design/env_fenestration_win.php

Whitestone Research (2008). *The Whitestone Building Maintenance and Repair Cost Reference 2008-2009*, 13th Annual Edition.

Wilson, J.B., Oregon State University (2008). CORRIM: Phase II Final Report, Module F, Particleboard: A Life-Cycle Inventory of Manufacturing Panels from Resource through Product. Found at http://www.corrim.org/pubs/reports/2010/phase2/Module_F.pdf.

Winiarski, D. et al. (2003) *The Business Case for Sustainable Design in Federal Facilities: Appendix B: Energy and Construction Cost Estimates*, Pacific Northwest National Laboratory and National Renewable Energy Laboratory.

World Steel Association (2011), Methodology report – Life cycle inventory study for steel products. http://www.worldsteel.org/

Appendix. Interpreting BIRDS Environmental Impact Scores: A Primer

Building ABC has a BIRDS Environmental Impact Score of 0.023 % and Building XYZ a score of 0.045 %. What does that mean?

Let's start from the beginning, considering just one building and one environmental impact at a time. Let's take a look, say, at the Global Warming performance of Building ABC, and ask:

Q. How much does the production and use of Building ABC over 40 years contribute to Global Warming?

A. BIRDS tells me that Building ABC contributes 8,520,000 kilograms of carbon dioxide-equivalent greenhouse gases over its life cycle.

Q. So what? All Buildings contribute greenhouse gases over their life cycle. Is 8,520,000 kilograms a lot or a little? How can I make sense of this number?

A. By relating the number to the total amount of greenhouse gases released every year in the United States. Let's name these annual U.S. releases "John Q. Public" and have him represent our yardstick. Now let's mark the spot showing Building ABC's greenhouse gases relative to his.

Q. Okay. Let's say you do that for Building ABC for all 12 environmental impacts. But then what? How can you combine all 12 yardsticks when they're measuring different things? Wouldn't you be mixing apples and oranges?

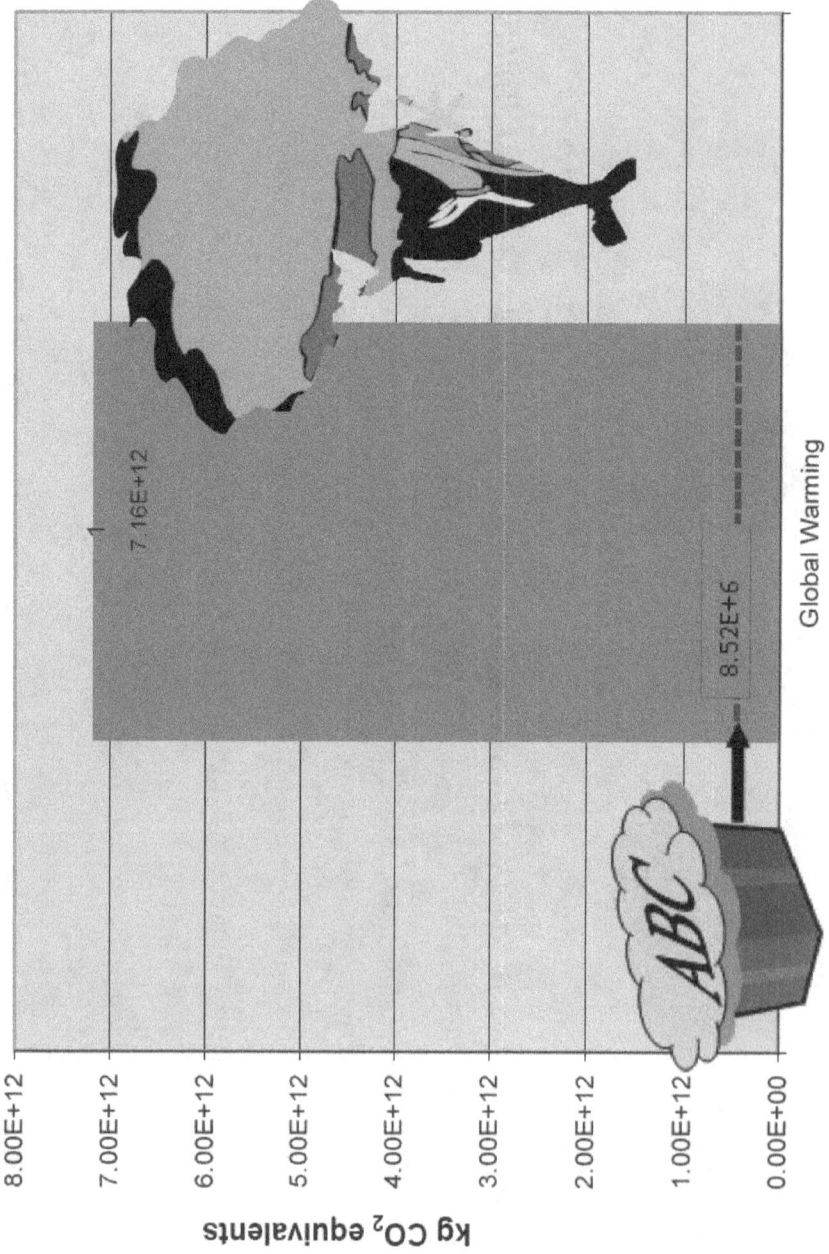

A. Yes, you would be, unless you made a single, common yardstick for all impacts—one based on Building ABC's *percentage share* of John Q. Public's impacts. That way, you could plot all impacts on the same graph. It's like a nutrition label, but instead of reporting a building's percentages of recommended daily allowances, we're reporting its percentages of John Q. Public's annual environmental impacts. Let's do this for Building ABC and Building XYZ.

Share of Annual
U.S. Impact

John Q. Public

5.10I

1.66E+12

4.64E+11 kg O3 eq

5.03E+05 CTUHnoncan

1.81E+09 acre

1.01E+10 kg N eq

3.82E+13 CTUE

1.69E+14 L

1.05E+04 CTUHcan

2.24E+10 kg PM10 eq

3.52E+13 kWh

7.16E+12 kg CO2 eq

100%
90%
80%
70%
60%
50%
40%
30%
20%
2.0E-6%
0%

ABC

Environmental Impact

- Global Warming
- Primary Energy Consumption
- HH Criteria Air
- HH Cancer
- Water Consumption
- Ecological Toxicity
- Eutrophication
- Land Use
- HH Noncancer
- Smog
- Acidification
- Ozone De

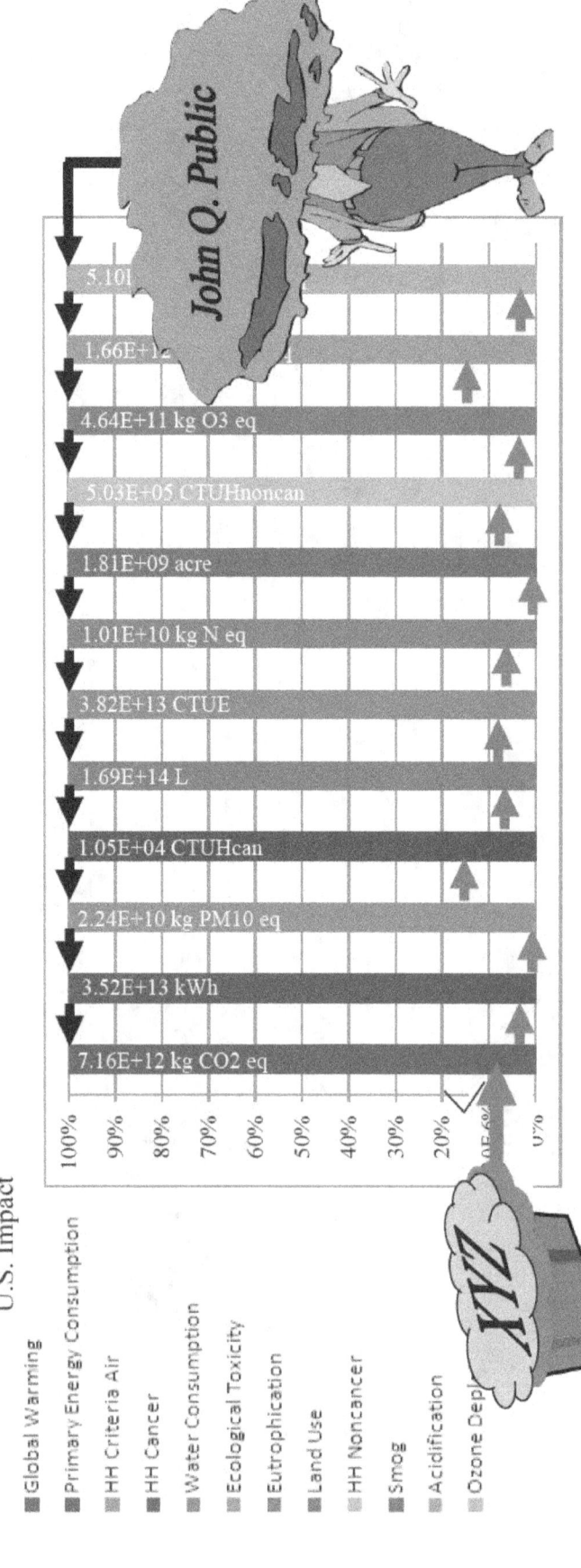

Share of Annual
U.S. Impact

Environmental Impact

Q. *I'm still confused. It looks like Building ABC scores better on Global Warming, but worse on Water Consumption, than Building XYZ. How do I know which Building is environmentally preferred, all things considered? Can't you just give me a simple average score?*

A. I could, but that would mean all environmental impacts are of the same importance. Most experts say that's not the case, so I'll give you a weighted average score instead. Then you can compare Building ABC side-by-side with Building XYZ when you're shopping for "green" Buildings. But always remember, it's better to have a *lower* BIRDS Environmental Impact Score. Think of the BIRDS Score as a penalty score—the higher it is, the worse it is.

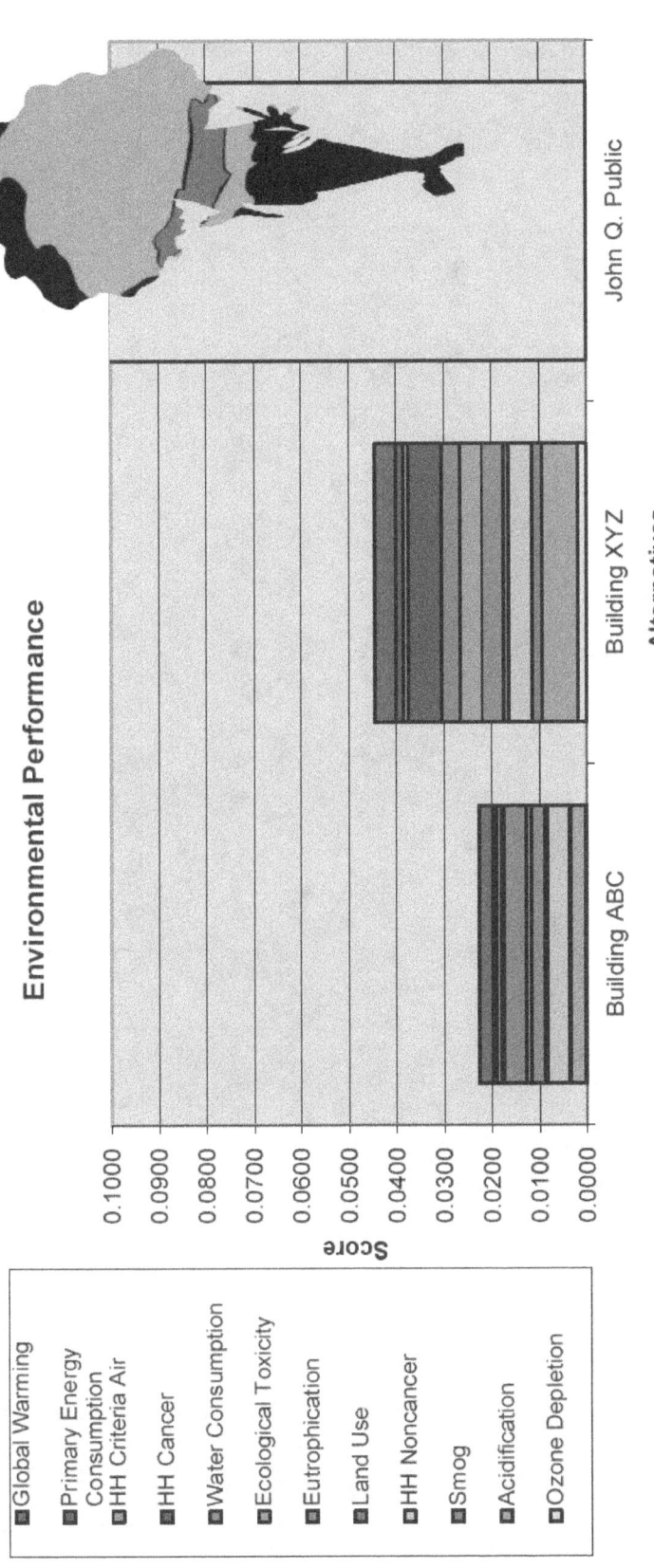

Environmental Performance

Q. Okay. But after all this, when I tell my colleagues that Building ABC, with a BIRDS Environmental Impact Score of 0.023 %, is greener than Building XYZ, with a score of 0.045 %, what am I really saying?

A. You're saying that, over its life cycle, Building ABC does less damage to the environment than does Building XYZ. If your colleague's eyes start to glaze over, quickly finish by saying that Buildings with lower BIRDS scores are greener. Otherwise, explain that Building ABC is greener because it contributes, on average, 0.023 % of annual U.S. environmental impacts, while Building XYZ contributes a larger share, 0.045 %.